Transmutation
Natural and Artificial

Nobel Prize Topics in Chemistry

A Series of Historical Monographs on
Fundamentals of Chemistry

Editor
Johannes W. van Spronsen
(The Hague and University of Utrecht)

Advisory Board
G. Dijkstra (Utrecht)   N. A. Figurowsky (Moscow)
F. Greenaway (London)   A. J. Ihde (Madison)
E. Rancke-Madsen (Copenhagen)   M. Sadoun-Goupil (Paris)
Irene Strube (Leipzig)   F. Szabadvary (Budapest)
T.J. Trenn (Munich)

Nobel Prize Topics in Chemistry traces the scientific development
of each subject for which a Nobel Prize was awarded in the light of
the historical, social and political background surrounding its
reception. In every volume one of the Laureate's most significant
publications is reproduced and discussed in the context of his life
and work, and the history of science in general.

This major series captures the intellectual fascination of a field
that is too often considered the domain of specialists, but which
nevertheless remains a significant area of study for all those
interested in the evolution of chemistry.

Current titles
Stereochemistry   *O. Bertrand Ramsay*
Transmutation: natural and artificial   *Thaddeus J. Trenn*
Inorganic Coordination Compounds   *George B. Kauffman*

Transmutation
Natural and Artificial

Thaddeus J. Trenn
Presently affiliated with the
Max Planck Institute of Physics, Munich, West Germany

LONDON · PHILADELPHIA · RHEINE

Heyden & Son Ltd, Spectrum House, Hillview Gardens, London NW4 2JQ, UK
Heyden & Son Inc., 247 South 41st Street, Philadelphia, PA 19104, USA
Heyden & Son GmbH, Devesburgstrasse 6, 4440 Rheine, West Germany

British Library Cataloguing in Publication Data

Trenn, Thaddeus J.
    Transmutation. – (Nobel prize topics in chemistry)
    1. Nuclear reactions – Addresses, essays, lectures
    I. Trenn, Thaddeus J.
    539.7'5   QC794.5

    ISBN 0-85501-685-X
    ISBN 0-85501-686-8 Pbk

Printed in Great Britain by Cambridge University Press, Cambridge

Contents

List of plates

Foreword

By Glenn T. Seaborg
(Nobel Laureate in Chemistry, 1951)

A member of the series of monographs Nobel Prize Topics in Chemistry, this book features three original published papers of winners of Nobel Prizes in the field of nuclear transmutation. Each of these three papers represents a milestone, a stage that opened extensive new territory for investigation at the time of publication. Each item of research involved a combination of the expertise of the chemist and the physicist.

First, representing the 'natural' aspect of the book's title *Transmutation – Natural and Artificial*, is the paper by New Zealand physicist Ernest Rutherford and English chemist Frederick Soddy, 'Radioactive Change', based on work done in Canada, which describes clearly for the first time the theory of transmutation as applied to the radioactive decay of members of the heavy natural uranium and thorium decay series. It is interesting to note their prescient observation that:

> the maintenance of solar energy, for example, no longer presents any fundamental difficulty if the internal energy of the component elements is considered to be available, i.e. if processes of sub-atomic change are going on.

The second paper, 'A New Type of Radioactivity' by the French wife–husband, chemist–physicist team, Irène Curie and Frédéric Joliot, represents 'artificial' transmutation induced by alpha particles projected from the 'natural' element polonium. They correctly foresaw that such 'lasting radioactivity' should also be obtained from bombardment with other particles, leading to the vast area of artificial radioactivity.

Finally, the famous publication by the German chemists, Otto Hahn and Fritz Strassmann, 'Concerning the Existence and Behavior of Alkaline Earth Metals Resulting from Neutron Irradiation of Uranium', unexpectedly revealed the discovery of the extraordinary 'artificial' nuclear transmutation which was soon interpreted as nuclear fission. These modest chemists included in their paper, following their tentative suggestion of this interpretation, the famous lines, ' ... we cannot bring ourselves yet to take such a drastic step which goes against all previous experience in nuclear physics'. However, they did immediately take this drastic step, as evidenced by their following confirmation experiments, buttressed by the observations of other scientists all over the world.

This monograph places these key discoveries in the field of nuclear transmutation in broad perspective by including a thorough description of historical background, contemporary impact and future expectations. Beginning with antiquity, the author guides us through alchemy, the rise of chemistry, the advent of the chemical atom and then into his theme, nuclear transmutation, starting at the beginning of this century and extending to the present. His concluding chapters deal with the extraordinarily important role that nuclear transmutation, through the fission reaction, plays as an indispensable, world-wide source of energy.

Glenn S. Seaborg

Professor of Chemistry,
University of California,
Berkeley, California, USA

General Editor's Preface

Nobel Prizes in Chemistry have been awarded almost every year since 1901, and the topics covered by these awards have touched upon almost every subject in chemistry. 'Nobel Prize Topics in Chemistry' plans to cover the history of each subject for which a Nobel Prize was awarded and to place particular emphasis on the life and work of the Nobel Prize winner himself. In this way the planned Series will come to describe the whole history of chemistry. The concept is to take one of each Nobel Laureate's most significant publications, to reprint it (as an English translation if appropriate), discuss it, and then to place it within the context of the Laureate's life and works in particular and the history of science in general, if possible going as far back as Egyptian, Babylonian and Greek antiquity. The Series will also look at possible future developments.

Contributions are presented in such a manner that a non-specialist background will suffice for the text to be comprehensible. The intention is to make as many readers as possible aware of and conversant with the problems underlying the development of various areas in the field of chemistry. Each volume attempts to give a smooth outline of the particular topic under consideration, uninterrupted by continual footnotes and references in the text.

The Series is not, in the first instance, aimed at the professional historian, but rather at the chemist, the research worker and the non-specialist who wishes to bring himself up to date on the historical background of one or more areas of chemistry. The student of chemistry and the historian or sociologist who for research wishes to focus broadly on one of the most spectacular disciplines in the natural sciences, can obtain a wide-ranging historical knowledge — knowledge which forms part of the general history of mankind and which can also be used to examine the reciprocal relationship between chemistry and society as a whole. The authors are well-known historians of chemistry or chemists with a solid knowledge of history. Among the latter group, occasionally a Nobel Prize winner will be our author. Nobel Prize winners are also contributing forewords to many of our volumes.

For certain parts of the texts a reasonable knowledge of chemistry – in some cases, the reading of formulas for example – is required, but the needs of the non-chemist have been anticipated by including in each volume a Glossary through which the reader can revise or extend his chemical knowledge. Each volume also contains a chronology of significant events and a detailed bibliography.

It is the aim of the Editor that the readers of this Series should obtain a clear idea of the particular experiences of a chemist in performing his research – research which sometimes led to a discovery of the greatest significance for humanity. It is thus intended not only to focus exclusively here on the main historical–chemical facts, but also to understand the chemist as a human being and look at the circumstances which led to his discoveries. However, to understand the social and eventually the political background of these historical developments, the reader must inform himself of the facts presented here. This Series thus aims to capture the intellectual fascination of a field that is too often considered to be the domain of specialists, but which nevertheless remains an area of proven intellectual adventure for all those who consider the quest for understanding the highest point to which man can aspire.

It is our sincere wish that the individual volumes in this Series shall realize these aims and intentions.

J. W. van SPRONSEN
The Hague

Author's Preface

In its millennial resilience, the speculation surrounding the transmutation of matter seems quite unrivalled in the entire history of science. All the more satisfying, then, was the ultimate justification of this belief and the perfection of this 'art'. The possibility of bringing the forces and ultimate constituents of nature under human control was always present, but any means by which to do so long remained unknown. Transmutation was first associated with the alchemical tradition dating back over two thousand years. Some sought to turn base metals into gold and others tried to secure a longer life. Whatever their goal, the alchemists sought to attain it with what became known vaguely as the 'philosophers' stone' or the 'elixir of life'. Three general features can be identified in the older alchemy: (1) that one substance was presumed to be transmutable into another; (2) that some principle of 'first matter' underlay such transmutation; and (3) that transmutation was an 'art' which could be performed by means of the philosophers' stone.

It turns out that nature has been adept at transmutation ever since the world began, but this fact was not discovered until the turn of the last century. The chemical elements are all composed of the same fundamental material units in varying arrays, and these arrangements can be transformed by interactions of certain types. Transmutations of this sort yield energy, not gold, but this only increases their value even more.

The first hint that nature was 'adept' came with the discovery of radioactivity in 1896. A few years later this was found to reflect a fundamental principle operative in nature (Nobel Prize Laureate 1908, Rutherford). It represents the final stage of a shaking out of residual instabilities inherent in the cosmic system. Over the next two decades, much was learned about the way this process works. Just twenty years later artificial transmutation was performed in the laboratory, disrupting nitrogen nuclei by means of 'alpha' projectiles provided through nature's own alchemy.

Protons were expelled in the process, and this was a new phenomenon not associated with natural transmutation. Perhaps protons were the long sought 'first matter'! That was 1917 and the results were announced in 1919. But the product of such a transformation need not be stable as was discovered in 1934 (Nobel Prize Laureate 1935, Joliot-Curie). The artificial radioactivity of such unstable products from alpha transmutation simulates nature's own removal of instabilities.

A feature common to both of these artificial processes, but absent in natural transmutation, was the existence of an intermediary stage whereby a proton might be expelled, although sometimes a neutron – the ultimate philosophers' stone! If alpha particles could disrupt nuclei, neutrons could do so even better. If the former could chip off bits, the latter could split it in two, as was discovered in 1938 (Nobel Prize Laureate 1944, Hahn). A fissioned nucleus, however, not only yields a great quantity of energy but also spews out, as it were, excess neutrons that are every bit as adept at fission as the former. These were soon harnessed in such a way that the resulting chain reaction could be made to release its vast energy either instantaneously or at a convenient rate. The new 'gold' was at hand!

Only after it had been learned how to perform this type of transmutation artificially – perhaps the most complex and difficult feat of transmutation yet achieved by man – did it become possible, paradoxically, to even recognize it in nature. One of man's technical wonders, the nuclear reactor, exists as a natural phenomenon! At least one such fossil reactor has already been discovered on the earth's surface. How many more are within the interior? As we continue to gain in understanding of the 'art' of transmutation, perhaps we will succeed in discovering and unraveling even more in the future about what is taking place in nature.

June 1981   THADDEUS J. TRENN
                                                                        Munich

Acknowledgements

Thanks are due to the following for permission to reproduce copyright material: Taylor & Francis Ltd for E. Rutherford and F. Soddy, 'Radioactive change', *Philosophical Magazine*, 6th series, **5**, 576–591 (1903), and for Plate 1 and Figs 11, 13 and 15, from T. J. Trenn, *The Self-Splitting Atom* (1978) and *Radioactivity and Atomic Change* (1975); the Académie des Sciences de l'Institut de France for the English translation of Irène Curie and F. Joliot, 'Un nouveau type de radioactivité', *Comptes Rendus de l'Académie des Sciences* **197**, 254–256 (1934); Springer Verlag, Dr Hans G. Graetzer and the *American Journal of Physics* for the English translation of O. Hahn and F. Strassman, 'Über den Nachweis und das Verhalten der bei der Bestrahlung des Urans mittels Neutronen entstehenden Erdalkalimetalle', *Naturwissenschaften* **27**, 11–15 (1939), in *American Journal of Physics* **32**, 9–15 (1964); the Division of Chemical Education, American Chemical Society, for Plates 2, 6 and 9, from M. E. Weeks, *Discovery of the Elements* (7th Edn, 1968); Mrs Dorothy Ainslie and Edward Arnold (Publishers) Ltd for Plate 7, from M. W. Travers, *A Life of Sir William Ramsay* (1956); University Books, Inc., for Fig. 3(a), from H. S. Redgrave, *Alchemy: Ancient and Modern* (1969; originally published by William Rider & Son Ltd, London, 1922); Bell & Hyman Ltd for Figs 3(b), 6 and 7(a), from J. Read, *Prelude to Chemistry* (1937); the Society for the History of Alchemy and Chemistry for Fig. 4, from H. J. Sheppard, *Ambix* **10**, 83–96 (1962); David Higham Associates Ltd and Henry Schuman, Inc., for Figs 5 and 7(b), from F. S. Taylor, *The Alchemists* (1949); the McGraw-Hill Book Company for Fig. 16, from R. D. Evans, *The Atomic Nucleus* (1955).

1

Adepts, awards and achievements

BIOGRAPHICAL INTRODUCTION

Transmutation has a very long history of failures but some important successes during the last century. In order to see where the historical path will lead, it is convenient to provide a reprint of the most important articles associated with each of the three Nobel Prize winners, a reflection on the significance that their achievements had for the development of chemistry, and a few biographical comments upon each of these adepts.

Transmutation is an important field of chemistry and physics, and so it is entirely fitting that several of the more prominent achievements and adepts* have been singled out to receive Nobel Prizes. No one was awarded the prize for the nuclear reactor or for the atomic bomb. But somewhat paradoxically the energy yield of the latter is often given in terms of the explosive power of so much dynamite, invented by Alfred Nobel.

The work for which the prizes were awarded was teamwork in all three cases. Ernest Rutherford (1871–1937) received the 1908 prize for providing the theory of atomic disintegration in 1903 worked out jointly with Frederick Soddy (1877–1956). Frédéric Joliot (1900–1958) and his wife Irène Joliot-Curie (1897–1956) received the 1935 prize jointly for their discovery of artificial radioactivity* in 1934. Otto Hahn (1879–1968) received the 1944 prize for providing definitive evidence with Lise Meitner (1878–1968) and Fritz Strassmann (1902–1980) of the fission* of uranium under neutron bombardment in 1938.

The Rutherford and Soddy paper of 1903 represents the culmination of nearly two years of collaborative research which led them inexorably to the disintegration theory as the proper theoretical explanation of radioactivity. According to this theory, radioactivity, as the emission of radioactive radiations, was a manifestation of a fundamental process taking place in nature independent of all known conditions. Atoms of

---

* Terms marked with an asterisk are defined in the Glossary (pp. 113–115).

matter — actually their nuclei* — were transmuting themselves into those
of different substances spontaneously, but at very well defined rates,
accompanied by the release of energy stored deep within the atoms. The
great significance for chemistry was clearly recognized at the time as reflected in the presentation speech.

> Rutherford's discoveries led to the highly surprising conclusion,
> that a chemical element . . . is capable of being transformed into
> other elements, and thus in a certain way it may be said that the
> progress of investigation is bringing us back once more to the
> transmutation theory propounded and upheld by the alchemists
> of old.
>
> As an explanation of these remarkable phenomena, Ruther-
> ford in conjunction with Dr F. Soddy . . . brought forward the
> so-called disintegration theory . . .
>
> According to this theory the origin and loss of radioactivity
> are to be regarded as due to changes — not in the molecule —
> but in the atom itself . . .
>
> The disintegration theory, and the experimental results upon
> which it is based, are synonymous with a new departure in
> chemistry, involving a fresh and decidedly extended comprehen-
> sion of the very basis of that science. To the chemists of the
> nineteenth century the atom and the element represented each
> in its sphere the uttermost limit of chemical subdivision or dis-
> integration, and at the same time the point beyond which it was
> impossible for experimental investigation to proceed . . . This
> line of demarcation, for so long regarded as insurmountable, has
> now been swept away. [Reference 1.†]

That was 1908 and Rutherford had become a scientific celebrity by that
time. He was born in New Zealand in 1871 and was a research student at
Cambridge, England from 1895. His first regular appointment was at
McGill in Montreal, where from 1901 he began a fruitful line of collaboration with the English chemist Soddy. Rutherford went to Manchester in
1907 and began investigations by which he identified the true nature of
alpha rays and utilized these particles as projectiles to explore the structure of the atom. His nucleus of 1911 was quickly developed by Bohr and
others into the nuclear model of the atom. Continuing to probe with
alpha particles*, Rutherford carried transmutation one step further during
the war with his discovery of artificial transmutation. His next position
was at Cambridge where from 1919 he pursued experiments in what
gradually became recognized as the field of nuclear physics. Rutherford
died suddenly in 1937 in his mid-60s and at the height of his powers and

† A list of the references cited will be found in Appendix C (p. 116).

Plate 1. Ernest Rutherford and Frederick Soddy at the time of their disintegration theory of radioactivity. (From T. J. Trenn, *The Self-Splitting
Atom*, frontispiece; courtesy Taylor & Francis Ltd.)

career. With the discovery of fission just a year later and the events which
immediately followed, Rutherford's beloved nuclear world had been
shaken to its very foundations. It would have been interesting to have had
his reaction.

The Joliot–Curie paper of 1934 was short, and entirely characteristic
of the publication tradition for contributions made to the French
Academy of Science. It made the point quite concisely that positive
electrons were being emitted in the manner of natural radioactivity, but
from non-radioactive substances which had been irradiated with alpha
particles and then had transmuted themselves from stable forms into
unstable ones after throwing off either a neutron* or a proton*. The
implication for chemistry was that instability could be introduced into
ordinary chemical substances. Radioactivity had been generally associated
only with the heaviest elements, and the only known exceptions were the
long-lived beta emitters, potassium ($^{40}_{19}K$) and rubidium ($^{87}_{37}Rb$), which
had been identified by 1908. But this new discovery made it possible for
any substance to become radioactive, and the general nature of the phenomenon became clearer. The common chemical elements were seen for
what they were, namely that which remains after nature's instabilities
are removed. And that this process is still taking place is evidenced by the

presence of the long lived radioactive substances uranium, thorium, rubidium and potassium.

Frédéric Joliot was born in Paris in 1900 and studied at the Ecole Supérieure from 1920 under the tutelage of Paul Langevin. From 1925 he worked under Madame Curie at the Institut du Radium, and in 1926 he married her daughter, Irène. From 1931 they began to collaborate on a series of researches, the first of which concerned the penetrating radiation noted by Bothe and Becker the previous year. In 1932 the Joliot-Curies announced that this penetrating radiation could kick protons out of substances, but they still construed the primary radiation as photons thereby missing the discovery of the neutron. As part of their research program on the emission of positrons by aluminum bombarded with alpha particles, Joliot covered the entrance to a cloud chamber* with a thin aluminum foil. The positive electrons induced in the foil by alpha particle bombardment did not cease to be registered upon removal of the polonium source. Repeating this result with a Geiger-Müller counter* to make sure that the secondary effect was not a function of the method of detection, they confirmed together the existence of artificial radioactivity and announced it early in 1934.

Plate 2. Jean-Frédéric Joliot and Irène Joliot-Curie. (From M. E. Weeks, *Discovery of the Elements*, pp. 804, 806; courtesy American Chemical Society.)

In 1937 Joliot received a professorship at the Collège de France but Irène continued her research at the Radium Institute. In 1937–38 she collaborated with Paul Savitch on experiments with neutron bombardment of uranium.* One of the products was similar to actinium but could be separated from it chemically like lanthanum. It *was* lanthanum, and fission had occurred, but the recognition of this fact came only later that year through the work of Hahn, Meitner and Strassmann. After the discovery of fission, Joliot immediately confirmed the high velocity of the fission products.* With Halban and Kowarski he also proved that uranium fission is accompanied by the emission of excess neutrons. Uranium bombarded with thermal neutrons yielded a flux of fast* neutrons which therefore could only have been a secondary emission. Perhaps even more important was the fact that on the average there were two or three secondary neutrons for every primary one. Excess neutrons provided the key to both atomic bombs and nuclear reactors. Joliot had planned the design of a 'pile' using heavy water as the moderator but the impending invasion of France (1940) prevented his group from conducting the experiment. After the war Joliot succeeded in greatly influencing the development of nuclear power in France. Irène died in 1956 of leukemia, *N B* like her mother, and Joliot died two years later.

The 1939 paper of Hahn and Strassmann is one of a series of papers associated with researches begun jointly with Lise Meitner. From 1934 Hahn and Meitner had been on the trail of the transuranic elements following Fermi's lead. But in 1938 Hahn and Strassmann found several more cases like the strange lanthanum of Irène Curie and Savitch; in particular they noted several with radium-like properties. Hahn concluded that neutron bombardment of uranium yields isotopes of radium. Meitner, an Austrian Jewess, meantime had had to leave Germany after the annexation of Austria that spring. Shortly after she arrived in Stockholm that July, she wrote to Hahn requesting irrefutable data in support of his claim. Strassmann and Hahn undertook a series of tests to prove that these products were chemically identical with radium. But they found instead that while these products could be precipitated with barium they could not be reseparated from barium. Hahn then reluctantly concurred with Strassmann's 'rash' suggestion that these products *were* isotopic varieties of barium and not radium at all.

Otto Hahn was born in Frankfurt in 1879 and studied organic chemistry. He spent the year 1904 in England working in Ramsay's London laboratory and became interested in radiochemistry. He showed a knack for finding previously unknown radioactive substances as exemplified by radiothorium and mesothorium. After a year's research in Montreal with Rutherford, Hahn returned to Berlin in 1907, and in that same year he began a thirty year collaboration with the physicist Lise Meitner. Hahn

Plate 3. (Above) Fritz Strassmann; (below) Lise Meitner and Otto Hahn. (Courtesy Fritz Strassmann.)

was one of the first to introduce the method of radioactive recoil in 1909, and in 1917 they discovered the parent of actinium. In 1921 he found the first example of nuclear isomerism – a fine structure within the isotopes. Their search for transuranic elements was a 'successful failure' yielding results they had not anticipated. He played no active role in the subsequent development of nuclear fission, although he retained an interest in the chemistry of the fission products. After the war Hahn became president of the Max-Planck-Gesellschaft and he remained active in science and politics until his death in 1968, three months before the death of Meitner.

The key publications of Rutherford, Joliot-Curie and Hahn – those most closely associated with that achievement for which each received the Nobel Prize – now follow.

RADIOACTIVE CHANGE

E. RUTHERFORD and F. SODDY

[from *Phil. Mag.* (6th ser.) 5, 576–591 (1903)]

1.   THE PRODUCTS OF RADIOACTIVE CHANGE AND THEIR SPECIFIC MATERIAL NATURE

In previous papers it has been shown that the radioactivity of the elements radium, thorium, and uranium is maintained by the continuous production of new kinds of matter which possess temporary activity. In some cases the new product exhibits well-defined chemical differences from the element producing it, and can be separated by chemical processes. Examples of this are to be found in the removal of thorium X from thorium and uranium X from uranium. In other cases the new products are gaseous in character, and so separate themselves by the mere process of diffusion, giving rise to the radioactive emanations which are produced by compounds of thorium and radium. These emanations can be condensed by cold and again volatilized; although they do not appear to possess positive chemical affinities, they are frequently occluded by the substances producing them when in the solid state, and are liberated by solution; they diffuse rapidly into the atmosphere and through porous partitions, and in general exhibit the behaviour of inert gases of fairly high molecular weight. In other cases again the new matter is itself non-volatile, but is produced by the further change of the gaseous emanation; so that the latter acts as the intermediary in the process of its separation from the radioactive element. This is the case with the two different kinds of excited activity produced on objects in the neighbourhood of compounds of thorium and radium respectively, which in turn possess well-defined and characteristic material properties. For example, the thorium excited activity is volatilized at a definite high temperature, and redeposited in the neighbourhood, and can be dissolved in some reagents and not in others.

These various new bodies differ from ordinary matter, therefore, only in one point, namely, that their quantity is far below the limit that can be reached by the ordinary methods of chemical and spectroscopic analysis. As an example that this is no argument against their specific material existence, it may be mentioned that the same is true of radium itself as it occurs in nature. No chemical or spectroscopic test is sufficiently delicate to detect radium in pitchblende, and it is not until the quantity present is increased many times by concentration that the characteristic spectrum begins to make its appearance. Mme. Curie and also Giesel have succeeded in obtaining quite considerable quantities of pure radium compounds by working up many tons of pitchblende, and the results go to show that radium is in reality one of the best defined and most characteristic of the chemical elements. So, also, the various new bodies, whose existence has been discovered by the aid of their radioactivity, would no doubt, like radium, be brought within the range of the older methods of investigation if it were possible to increase the quantity of material employed indefinitely.

2.   THE SYNCHRONISM BETWEEN THE CHANGE AND THE RADIATION

In the present paper the nature of the changes in which these new bodies are produced remains to be considered. The experimental evidence that has been accumulated is now sufficiently complete to enable a general theory of the nature of the process to be established with a considerable degree of certainty and definiteness. It soon became apparent from this evidence that a much more intimate connexion exists between the radioactivity and the changes that maintain it than is expressed in the idea of the production of active matter. It will be recalled that all cases of radioactive change that have been studied can be resolved into the production by one substance of one other (disregarding for the present the expelled rays). When several changes occur together these are not simultaneous but successive. Thus thorium produces thorium X, the thorium X produces the thorium emanation, and the latter produces the excited activity. Now the radioactivity of each of these substances can be shown to be connected, not with the change in which it was itself produced, but with the change in which it in turn produces the next new type. Thus after thorium X has been separated from the thorium producing it, the radiations of the thorium X are proportional to the amount of emanation that it produces, and both the radioactivity and the emanating power of thorium X decay according to the same law *and at the same rate.* In the next stage the emanation goes on to produce the excited activity. The activity of the emanation falls to half-value in one minute, and the amount of excited activity produced by it on the negative electrode in an electric field falls off in like ratio. These results are fully borne out in the case of radium. The activity of the radium emanation decays to half-value in four days, and so also does its power of producing the excited activity.

Hence it is not possible to regard radioactivity as a *consequence* of changes that have already taken place. The rays emitted must be an *accompaniment* of the change of the radiating system into the one next produced.

Non-separable activity. This point of view at once accounts for the existence of a constant radioactivity, non-separable by chemical processes, in each of the three radio-elements. This non-separable activity consists of the radiations that accompany the primary change of the radio-element itself into the first new product that is produced. Thus in thorium about 25 per cent of the α radiation accompanies the first change of the thorium into thorium X. In uranium the whole of the α radiation is non-separable and accompanies the change of the uranium into uranium X.

Several important consequences follow from the conclusion that the radiations accompany the change. A body that is radioactive must *ipso facto* be changing, and hence it is not possible that any of the new types of radioactive matter — e.g., uranium X, thorium X, the two emanations, &c. — can be identical with any of the known elements. For they remain in existence only a short time, and the decay of their radio-activity is the expression of their continuously diminishing quantity. On the other hand, since the ultimate products of the changes cannot be radioactive, there must always exist at least one stage in the process beyond the range of the methods of experiment. For this reason the ultimate products that result from the changes remain unknown, the quantities involved being unrecognizable, except by the methods of radioactivity. In the naturally occurring minerals containing the radio-elements these changes must have been proceeding steadily over very long periods, and, unless they succeed in escaping, the ultimate products should have accumulated in sufficient quantity to be detected, and therefore should appear in nature as the invariable com-panions of the radio-elements. We have already suggested on these and other grounds that possibly helium may be such an ultimate product, although, of course, the suggestion is at present a purely speculative one. But a closer study of the radioactive minerals would in all probability afford further evidence on this important question.

3. THE MATERIAL NATURE OF THE RADIATIONS

The view that the ray or rays from any system are produced at the moment the system changes has received strong confirmation by the discovery of the electric and magnetic deviability of the α ray. The deviation is in the opposite sense to the β or cathode-ray, and the rays thus consist of positively charged bodies projected with great velocity (Rutherford, Phil. Mag., Feb. 1903). The latter was shown to be of the order of 2.5×10^9 cms. per second. The value of e/m, the ratio of the charge of the carrier to its mass, is of the order 6×10^3. Now the value of e/m for the cathode-ray is about 10^7. Assuming that the value of the charge is the same in each case, the apparent mass of the positive projected particle is over 1000 times as great as for the cathode-ray. Now $e/m = 10^4$ for the hydrogen atom in the electrolysis of water. The particle that constitutes the α ray thus behaves as if its mass were of the same order as that of the hydrogen atom. The α rays from all the radio-elements, and from the various radioactive bodies which they produce, possess analogous properties, and differ only to a slight extent in penetrating power. There are thus strong reasons for the belief that the α rays generally are projections and that the mass of the particle is of the same order as that of the hydrogen atom, and very large compared with the mass of the projected particle which constitutes the β or easily deviable ray from the same element.

With regard to the part played in radioactivity by the two types of radiation, there can be no doubt that the α rays are by far the more important. In all cases they represent over 99 per cent of the energy radiated*, and although the β rays on account of their penetrating power and marked photographic action have been more often studied, they are comparatively of much less significance.

It has been shown that the non-separable activity of all three radio-elements, the activity of the two emanations, and the first stage of the excited activity of radium, comprise only α rays. It is not until the processes near completion in so far as their

---

* In the paper in which this is deduced (Phil. Mag. Sept. 1902, p. 329) there is an obvious slip of calculation. The number there should be 100 instead of 1000, hence yielding 90% instead of 99%.

progress can be experimentally traced that the β or cathode-ray makes its appearance.[†]

In light of this evidence there is every reason to suppose, not merely that the expulsion of a charged particle accompanies the change, but that this expulsion actually *is* the change.

4.   THE LAW OF RADIOACTIVE CHANGE

The view that the radiation from an active substance accompanies the change gives a very definite physical meaning to the law of decay of radioactivity. In all cases where one of the radioactive products has been separated, and its activity examined independently of the active substance which gives rise to it, or which it in turn produces, it has been found that the activity under all conditions investigated falls off in a geometrical progression with the time. This is expressed by the equation

$$\frac{I_t}{I_o} = \epsilon^{-\lambda t}$$

where I_o is the initial ionization current due to the radiations, I_t that after the time t, and λ is a constant. Each ray or projected particle will in general produce a certain definite number of ions in its path, and the ionization current is therefore proportional to the number of such particles projected per second. Thus

$$\frac{n_t}{n_o} = \epsilon^{-\lambda t},$$

where n_t is the number projected in unit of time for the time t and n_o the number initially.

If each changing system gives rise to one ray, the number of systems N_t which remain unchanged at the time t is given by

$$N_t = \int_t^\infty n_t \cdot dt = \frac{n_o}{\lambda} \epsilon^{-\lambda t}.$$

The number N_o initially present is given by putting $t = 0$.

$$N_o = \frac{n_o}{\lambda}$$

and

$$\frac{N_t}{N_o} = \epsilon^{-\lambda t}.$$

The same law holds if each changing system produces two or any definite number of rays.

† In addition to the α and β rays the radio-elements also give out a third type of radiation which is extremely penetrating. Thorium as well as radium (Rutherford, *Phys. Zeit.* 1902) gives out these penetrating rays, and it has since been found that uranium possesses the same property. These rays have not yet been sufficiently examined to make any discussion possible of the part they play in radioactive processes.

Differentiating

$$\frac{d\text{N}}{dt} = -\lambda \text{N}_t \, ,$$

or, the rate of change of the system at any time is always proportional to the amount remaining unchanged.

The law of radioactive change may therefore be expressed in the one statement — the proportional amount of radioactive matter that changes in unit time is a constant. When the total amount does not vary (a condition nearly fulfilled at the equilibrium point where the rate of supply is equal to the rate of change) the proportion of the whole which changes in unit time is represented by the constant λ, which possesses for each type of active matter a fixed and characteristic value. λ may therefore be suitably called the "radioactive constant". The complexity of the phenomena of radioactivity is due to the existence as a general rule of several different types of matter changing at the same time into one another, each type possessing a different radioactive constant.

5.   THE CONSERVATION OF RADIOACTIVITY

The law of radioactive change that has been deduced holds for each stage that has been examined, and therefore holds for the phenomenon generally. The radioactive constant λ has been investigated under very widely varied conditions of temperature, and under the influence of the most powerful chemical and physical agencies, and no alteration of its value has been observed. The law forms in fact the mathematical expression of a general principle to which we have been led as the result of our investigations as a whole. Radioactivity, according to present knowledge, must be regarded as the result of a process which lies wholly outside the sphere of known controllable forces, and cannot be created, altered, or destroyed. Like gravitation, it is proportional only to the quantity of matter involved, and in this restricted sense it is therefore true to speak of the principle as the conservation of radioactivity*. Radioactivity differs of course from gravitation in being a special and not necessarily a universal property of matter, which is possessed by different kinds in widely different degree. In the processes of radioactivity these different kinds change into one another and into inactive

---

*   Apart from the considerations that follow, this nomenclature is a convenient expression of the observed facts that the total radioactivity (measured by the radiations peculiar to the radio-elements) is for any given mass of radio-element a constant under all conditions investigated. The radioactive equilibrium may be disturbed and the activity distributed among one or more active products capable of separation from the original element. But the sum total throughout these operations is at all times the same.

For practical purposes the expression "conservation", applied to the radioactivity of the three radio-elements, is justified by the extremely minute proportion that can change in any interval over which it is possible to extend actual observations. But *rigidly* the term "conservation" applies only with reference to the radioactivity of any definite quantity of radioactive matter, whereas in nature this quantity must be changing spontaneously and continually growing less. To avoid possible misunderstanding, therefore, it is necessary to use the expression only in this restricted sense.

matter, producing corresponding changes in the radioactivity. Thus the decay of radio-activity is to be ascribed to the disappearance of the active matter, and the recovery of radioactivity to its production. When the two processes balance — a condition very nearly fulfilled in the case of the radio-elements in a closed space — the activity re-mains constant. But here the apparent constancy is merely the expression of the slow rate of change of the radio-element itself. Over sufficiently long periods its radioactiv-ity must also decay according to the law of radioactive change, for otherwise it would be necessary to look upon radioactive change as involving the creation of matter. In the universe therefore the total radioactivity must, according to our present know-ledge, be growing less and tending to disappear. Hence the energy liberated in radio-active processes does not disobey the law of the conservation of energy.

It is not implied in this view that radioactivity, considered with reference to the quantity of matter involved, is conserved under all conceivable conditions, or that it will not ultimately be found possible to control the processes that give rise to it. The principle enunciated applies of course only to our present state of experimental know-ledge, which is satisfactorily interpreted by its aid.

The general evidence on which the principle is based embraces the whole field of radioactivity. The experiments of Becquerel and Curie have shown that the radiations from uranium and radium respectively remain constant over long intervals of time. Mme. Curie put forward the view that radioactivity was a specific property of the element in question, and the successful separation of the element radium from pitch-blende was a direct result of this method of regarding the property. The possibility of separating from a radio-element an intensely active constituent, although at first sight contradictory, has afforded under closer examination nothing but confirmation of this view. In all cases only a part of the activity is removed, and this part is recovered spontaneously by the radio-element in the course of time. Mme. Curie's original posi-tion, that radioactivity is a specific property of the element, must be considered to be beyond question. Even if it should ultimately be found that uranium and thorium are admixtures of these elements with a small *constant* proportion of new radio-elements with correspondingly intense activity, the general method of regarding the subject is quite unaffected.

In the next place, throughout the course of our investigations we have not observed a single instance in which radioactivity has been created in an element not radioactive, or destroyed or altered in one that is, and there is no case at present on record in which such a creation or destruction can be considered as established. It will be shown later that radioactive change can only be of the nature of an atomic dis-integration, and hence this result is to be expected, from the universal experience of chemistry in failing to transform the elements. For the same reason it is not to be expected that the rate of radioactive change would be affected by known physical or chemical influences. Lastly, the principle of the conservation of radioactivity is in agreement with the energy relations of radioactive change. These will be considered more fully in § 7, where it is shown that the energy changes involved are of a much higher order of magnitude than is the case in molecular change.

It is necessary to consider briefly some of the apparent exceptions to this principle of the conservation of radioactivity. In the first place it will be recalled that the emanating power of the various compounds of thorium and radium respectively differ widely among themselves, and are greatly influenced by alterations of physical state.

It was recently proved (Phil. Mag. April 1903, p. 453) that these variations are caused by alterations in the rate at which the emanations escape into the surrounding atmosphere. The emanation is produced at the same rate both in de-emanated and in highly emanating thorium and radium compounds, but is in the former stored up or occluded in the compound. By comparing the amount stored up with the amount produced per second by the same compound dissolved, it was found possible to put the matter to a very sharp experimental test which completely established the law of the conservation of radioactivity in these cases. Another exception is the apparent destruction of the thorium excited activity deposited on a platinum wire by ignition to a white heat. This has recently been examined in this laboratory by Miss Gates, and it was found that the excited activity is not destroyed, but is volatilized at a definite temperature and redeposited in unchanged amount on the neighbouring surfaces.

Radioactive "Induction". Various workers in this subject have explained the results they have obtained on the idea of radioactive "induction", in which a radioactive substance has been attributed the power of inducing activity in bodies mixed with it, or in its neighbourhood, which are not otherwise radioactive. This theory was put forward by Becquerel to explain the fact that certain precipitates (notably barium sulphate) formed in solutions of radioactive salts are themselves radioactive. The explanation has been of great utility in accounting for the numerous examples of the presence of radioactivity in non-active elements, without the necessity of assuming in each case the existence of a new radio-element therein, but our own results do not allow us to accept it.

In the great majority of instances that have been recorded the results seem to be due simply to the *mixture of active matter with the inactive element.* In some cases the effect is due to the presence of a small quantity of the original radio-element, in which case the "induced" activity is permanent. In other cases, one of the disintegration products, like uranium X or thorium X, has been dragged down by the precipitate, producing temporary, or, as it is sometimes termed, "false" activity. In neither case is the original character of the radiation at all affected. It is probable that a re-examination of some of the effects that have been attributed to radioactive induction would lead to new disintegration products of the known radio-elements being recognized.

Other Results. A number of cases remain for consideration, where, by working with very large quantities of material, there have been separated from minerals possible new radio-elements, i.e. substances possessing apparently permanent radioactivity with chemical properties different from those of the three known radio-elements. In most of these cases, unfortunately, the real criteria that are of value, viz., the nature of the radiations and the presence or absence of distinctive emanations, have not been investigated. The chemical properties are of less service, for even if a new element were present, it is not at all necessary that it should be in sufficient quantity to be detected by chemical or spectroscopic analysis. Thus the radio-lead described by Hoffmann and Strauss and by Giesel cannot be regarded as a new element until it is shown that it has permanent activity of a distinctive character.

In this connexion the question whether polonium (radio-bismuth) is a new element is of great interest. The polonium discovered by Mme. Curie is not a permanent radioactive substance, its activity decaying slowly with the time. On the view put forward in these papers, polonium must be regarded as a disintegration product of one of the radio-elements present in pitchblende. Recently, however, Marckwald (*Ber. der D. Chem. Gesel.* 1902, pp. 2285 & 4239), by the electrolysis of pitchblende solutions, has obtained an intensely radioactive substance very analogous to the polonium of Curie. But he states that the activity of his preparation does not decay with time, and this, if confirmed, is sufficient to warrant the conclusion that he is not dealing with the same substance as Mme. Curie. On the other hand, both preparations give only α rays, and in this they are quite distinct from the other radio-elements. Marckwald has succeeded in separating his substance from bismuth, thus showing it to possess different chemical properties, and in his latest paper states that the bismuth-free product is indistinguishable chemically from tellurium. If the permanence of the radioactivity is established, the existence of a new radio-element must be inferred.

If elements heavier than uranium exist it is probable that they will be radioactive. The extreme delicacy of radioactivity as a means of chemical analysis would enable such elements to be recognized even if present in infinitesimal quantity. It is therefore to be expected that the number of radio-elements will be augmented in the future, and that considerably more than the three at present recognized exist in minute quantity. In the first stage of the search for such elements a purely chemical examination is of little service. The main criteria are the permanence of the radiations, their distinctive character, and the existence or absence of distinctive emanations or other disintegration products.

6.   THE RELATION OF RADIOACTIVE CHANGE TO CHEMICAL CHANGE

The law of radioactive change, that the rate of change is proportional to the quantity of changing substance, is also the law of monomolecular chemical reaction. Radioactive change, therefore, must be of such a kind as to involve one system only, for if it were anything of the nature of a combination, where the mutual action of two systems was involved, the rate of change would be dependent on the concentration, and the law would involve a volume-factor. This is not the case. Since radioactivity is a specific property of the element, the changing system must be the chemical atom, and since only one system is involved in the production of a new system and, in addition, heavy charged particles, in radioactive change the chemical atom must suffer disintegration.

The radio-elements possess of all elements the heaviest atomic weight. This is indeed their sole common chemical characteristic. The disintegration of the atom and the expulsion of heavy charged particles of the same order of mass as the hydrogen atom leaves behind a new system lighter than before, and possessing chemical and physical properties quite different from those of the original element. The disintegration process, once started, proceeds from stage to stage with definite measurable velocities in each case. At each stage one or more α "rays" are projected, until the last stages are reached, when the β "ray" or electron is expelled. It seems advisable to possess a special name for these now numerous atom-fragments, or new atoms, which result from the original atom after the ray has been expelled, and which remain in existence only a limited time, continually undergoing further change. Their instability is their chief characteristic. On the one hand, it prevents the quantity accumulating, and in consequence it is hardly likely that they can ever be investigated by the ordinary methods. On the other, the instability and consequent ray-expulsion furnishes the means whereby they can be investigated. We would therefore suggest the term *metabolon* for this purpose. Thus in the following table the metabolons at present known to result from the disintegration of the three radio-elements have been arranged in order.

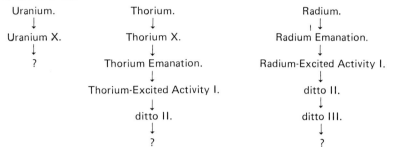

Uranium.   Thorium.   Radium.
↓   ↓   ↓
Uranium X.   Thorium X.   Radium Emanation.
↓   ↓   ↓
?   Thorium Emanation.   Radium-Excited Activity I.
                            ↓   ↓
                            Thorium-Excited Activity I. ditto II.
                            ↓   ↓
                            ditto II.   ditto III.
                            ↓   ↓
                            ?   ?

The three queries represent the three unknown ultimate products. The atoms of the radio-elements themselves form, so to speak, the common ground between metabolons and atoms, possessing the properties of both. Thus, although they are disintegrating, the rate is so slow that sufficient quantity can be accumulated to be investigated chemically. Since the rate of disintegration is probably a million times faster for radium than it is for thorium or uranium, we have an explanation of the excessively minute proportion of radium in the natural minerals. Indeed, every consideration points to the conclusion that the radium atom is also a metabolon in the full sense of having been formed by disintegration of one of the other elements present in the mineral. For example, an estimation of its "life", goes to show that the latter can hardly be more than a few thousand years (see § 7). The point is under experimental investigation by one of us, and a fuller discussion is reserved until later.

There is at present no evidence that a single atom or metabolon ever produces more than one new kind of metabolon at each change, and there are no means at present of finding, for example, either how many metabolons of thorium X, or how many projected particles, or "rays", are produced from each atom of thorium. The simplest plan therefore, since it involves no possibility of serious error if the nature of the convention is understood, is to assume that each atom or metabolon produces one new metabolon or atom and one "ray".

7.   THE ENERGY OF RADIOACTIVE CHANGE, AND THE INTERNAL
      ENERGY OF THE CHEMICAL ATOM

The position of the chemical atom as a very definite stage in the complexity of matter, although not the lowest of which it is now possible to obtain experimental knowledge, is brought out most clearly by a comparison of the respective energy relations of radioactive and chemical change. It is possible to calculate the order of the quantity of energy radiated from a given quantity of radio-element during its complete change, by several independent methods, the conclusions of which agree very well among themselves. The most direct way is from the energy of the particle projected, and the total number of atoms. For each atom cannot produce less than one "ray" for each change it undergoes, and we therefore arrive in this manner at a minimum estimate of the total energy radiated. On the other hand, one atom of a radio-element, if completely resolved into projected particles, could not produce more than about 200 such particles at most, assuming that the mass of the products is equal to the mass of the atom. This consideration enables us to set a maximum limit to the estimate. The α rays represent so large a proportion of the total energy of radiation that they alone need be considered.

Let m = mass of the projected particle,
v = the velocity,
e = charge.
Now for the α ray of radium
$$v = 2.5 \times 10^9,$$
$$e/m = 6 \times 10^3.$$
The kinetic energy of each particle

$$\tfrac{1}{2}mv^2 = \frac{1m}{2e} v^2 e = 5 \times 10^{14} e.$$

J. J. Thomson has shown that
$$e = 6 \times 10^{-10} \text{ E.S. Units} = 2 \times 10^{-20} \text{ Electromagnetic Units.}$$

Therefore the kinetic energy of each projected particle = 10^{-5} erg. Taking 10^{20} as the probable number of atoms in one gram of radium, the total energy of the rays from the latter = 10^{15} ergs = 2.4×10^7 gram-calories, on the assumption that each atom projects one ray. Five successive stages in the disintegration are known, and each stage corresponds to the projection of at least one ray. It may therefore be stated that the total energy of radiation during the disintegration of one gram of radium cannot be less than 10^8 gram-calories, and may be between 10^9 and 10^{10} gram-calories. The energy radiated does not necessarily involve the whole of the energy of disintegration and may be only a small part of it. 10^8 gram-calories per gram may therefore be safely accepted as the least possible estimate of the energy of radioactive change in radium. The union of hydrogen and oxygen liberates approximately 4×10^3 gram-calories per gram of water produced, and this reaction sets free more energy for a given weight than any other chemical change known. The energy of radioactive change must therefore be at least twenty-thousand times, and may be a million times, as great as the energy of any molecular change.

The rate at which this store of energy is radiated, and in consequence the life of a radio-element, can now be considered. The order of the total quantity of energy liberated per second in the form of rays from 1 gram of radium may be calculated from the total number of ions produced and the energy required to produce an ion. In the solid salt a great proportion of the radiation is absorbed in the material, but the difficulty may be to a large extent avoided by determining the number of ions produced by the radiation of the emanation, and the proportionate amount of the total radiation of radium due to the emanation. In this case most of the rays are absorbed in producing ions from the air. It was experimentally found that the maximum current due to the emanation from 1 gram of radium, of activity 1000 compared with uranium, in a large cylinder filled with air, was 1.65×10^{-8} electromagnetic units.

Taking $e = 2 \times 10^{-20}$, the number of ions produced per second $= 8.2 \times 10^{11}$. These ions result from the collision of the projected particles with the gas in their path. Townsend (Phil. Mag. 1901, vol. i.), from experiments on the production of ions by collision, has found that the minimum energy required to produce an ion is 10^{-11} ergs. Taking the activity of pure radium as a million times that of uranium, the total energy radiated per second by the emanation from 1 gram of pure radium = 8200 ergs. In radium compounds in the solid state, this amount is about 0.4 of the total energy of radiation, which therefore is about

$$2 \times 10^4 \text{ ergs per second,}$$
$$6,3 \times 10^{11} \text{ ergs per year,}$$
$$15,000 \text{ gram-calories per year.}$$

This again is an under-estimate, for only the energy employed in producing ions has been considered, and this may be only a small fraction of the total energy of the rays.

Since the α radiation of all the radio-elements is extremely similar in character, it appears reasonable to assume that the feebler radiations of thorium and uranium are due to these elements disintegrating less rapidly than radium. The energy radiated in these cases is about 10^{-6} that from radium, and is therefore about .015 gram-calorie per year. Dividing this quantity by the total energy of radiation, 2.4×10^7 gram-calories, we obtain the number 6×10^{-10} as a maximum estimate for the proportionate amount of uranium or thorium undergoing change per year. Hence in one gram of these elements less than a milligram would change in a million years. In the case of radium, however, the same amount must be changing per gram *per year*. The "life" of the radium cannot be in consequence more than a few thousand years on this minimum estimate, based on the assumption that each particle produces one ray at each change. If more are produced the life becomes correspondingly longer, but as a maximum the estimate can hardly be increased more than 50 times. So that it appears certain that the radium present in a mineral has not been in existence as long as the mineral itself, but is being continually produced by radioactive change.

Lastly, the number of "rays" produced per second from 1 gram of a radio-element may be estimated. Since the energy of each "ray" $= 10^{-5}$ ergs $= 2.4 \times 10^{-13}$ gram-calories, 6×10^{10} rays are projected every year from 1 gram of uranium. This is approximately 2000 per second. The α radiation of 1 milligram of uranium in one second is probably within the range of detection by the electrical method. The methods of experiment are therefore almost equal to the investigation of a single atom disintegrating, whereas not less than 10^4 atoms of uranium could be detected by the balance.

It has been pointed out that these estimates are concerned with the energy of radiation, and not with the total energy of radioactive change. The latter, in turn, can only be a portion of the internal energy of the atom, for the internal energy of the resulting products remains unknown. All these considerations point to the conclusion that the energy latent in the atom must be enormous compared with that rendered free in ordinary chemical change. Now the radio-elements differ in no way from the other elements in their chemical and physical behaviour. On the one hand they resemble chemically their inactive prototypes in the periodic system very closely, and on the other they possess no common chemical characteristic which could be associated with their radioactivity. Hence there is no reason to assume that this enormous store of energy is possessed by the radio-elements alone. It seems probable that atomic energy in general is of a similar, high order of magnitude, although the absence of change prevents its existence being manifested. The existence of this energy accounts for the stability of the chemical elements as well as for the conservation of radio-activity under the influence of the most varied conditions. It must be taken into account in cosmical physics. The maintenance of solar energy, for example, no longer presents any fundamental difficulty if the internal energy of the component elements is considered to be available, i.e. if processes of sub-atomic change are going on. It is interesting to note that Sir Norman Lockyer has interpreted the results of his spectro-scopic researches on the latter view (Inorganic Evolution, 1900) although he regards the temperature as the cause rather than the effect of the process.

McGill University
Montreal

A NEW TYPE OF RADIOACTIVITY

Note by Mme IRENE CURIE and M.F. JOLIOT
presented by M. Jean Perrin

[translated from 'Un nouveau type de radioactivité',
C.R. Acad. Sci. (Paris) 197, 254—256 (1934)]

We have shown recently by means of the Wilson method[1] that certain light elements (glucinium [beryllium], boron, aluminium) emit positive electrons when irradiated by α-rays of polonium. According to our interpretation, the emission of positive electrons by Be may be due to *internal materialization* of γ-rays, while the positive electrons emitted by B and Al could be *transmutation electrons* associated with the emission of neutrons.

While looking for the correct interpretation of the mechanism causing these emissions, we discovered the following phenomenon:

Emission of positive electrons by certain light elements irradiated by α-rays of polonium, after the source of α-rays is removed, continues for more or less of a long time, and in the case of boron for more than half an hour.

We placed an aluminium foil at a distance of 1 mm from the source of polonium. After the foil was irradiated for approximately 10 min, we placed it on top of a Geiger-Müller counter fitted with an aluminium screen 0.07 mm thick which covered the aperture. We observed that the foil emitted radiation, the intensity of which decreases exponentially with time, with a period of 3 min 15 s. A similar result is obtained with boron and magnesium but with *different* periods of decay, 14 min for boron and 2 min 30 s for magnesium.

The intensity of radiation (immediately after exposure to α-rays) increases with the time of irradiation up to a certain limit. In this way, initial intensities of the same order are obtained for elements B, Mg, Al, about 150 pulses/min in the counter, when using a 60 millicurie source of polonium.

No such effect was observed with elements H, Li, C, Be, N, O, F, Na, Ca, Ni, Ag.[2] For some elements the phenomenon probably does not occur, for others the period of decay is perhaps too short.

Experiments carried out either with the Wilson method or with the 'trochoid' [excited spectrum 'focusing'] method introduced by Thibaud have shown that radiation emitted by boron and aluminium consists of positive electrons. Probably the same goes for the radiation of magnesium.

By introducing a copper screen between the counter and the irradiated foil, it has been established that in the case of Al the major part of radiation was absorbed in 0.88 g/cm^2, and in the case of B and Mg, in 0.26 g/cm^2, where the corresponding energies would be 2.2 $\times$ 10^6 eV for Al, and 0.7 $\times$ 10^6 eV for B and Mg, assuming that the same law of absorption applies as for negative electrons.

When the energy of α-rays irradiating the aluminium is reduced, the number of positive electrons diminishes, although the period of decay does not seem to change. Almost no positive electrons are observed when the energy of α-rays falls below 10^6 eV.

These experiments show the existence of a new type of radioactivity with the emission of positive electrons. We think that the process of emission for aluminium could be described as follows:

$$_{13}^{27}\text{Al} + {}_2^4\text{He} = {}_{15}^{30}\text{P} + {}_0^1n$$

The $_{15}^{30}$P isotope of phosphorus should be radioactive with a period of 3 min 15 s, emitting positive electrons according to the following reaction:

$$_{15}^{30}\text{P} = {}_{14}^{30}\text{Si} + e^+$$

An analogous reaction could be expected for boron and magnesium, the unstable nuclei being $_7^{13}$N and $_{14}^{27}$Si. The isotopes of $_7^{13}$N, $_{14}^{27}$Si and $_{15}^{30}$P can exist only for a rather short period of time, and this is why they would not be observed in nature.

In our view the explanation according to which

$$_{13}^{27}\text{Al} + {}_2^4\text{He} = {}_{14}^{30}\text{Si} + {}_1^1\text{H}$$

$$_{14}^{30}\text{Si} = {}_{14}^{30}\text{Si} + e^+ + e^-$$

is not very likely, as the excited isotope of $_{14}^{30}$Si able to become deactivated in the course of time would have sufficient energy to create a pair of electrons. No emission of negative electrons is observed, and theoretically it is very improbable that the energy differential between the electrons would be sufficient to prevent the negatives from being observed.[3] On the other hand, it can be assumed that in this process the duration of the excitation state would be unusually long with a coefficient of internal materialization equal to unity.

In effect, it was possible for the first time to create, with the aid of an external agent, radioactivity in certain atomic nuclei lasting a measurable time in the absence of the source of excitation.

A lasting radioactivity similar to that observed by us can undoubtedly occur in the case of bombardment by other particles. The same radioactive atom could surely be created through several nuclear reactions. For example, the nucleus $_7^{13}$N, which according to our hypothesis is radioactive, could be obtained by the interaction of a deuteron with carbon after emission of a neutron.

NOTES

[1] Comptes rendus, 196, 1933, p. 1885; J. de Phys. et Rad., 4, 1933, p. 494.
[2] Therefore this phenomenon cannot be due to contamination of the source of polonium.
[3] Nedelsky and Oppenheimer, Phys. Rev., 44, 1933, p. 948.

CONCERNING THE EXISTENCE AND BEHAVIOR OF ALKALINE EARTH METALS RESULTING FROM NEUTRON IRRADIATION OF URANIUM[1]

O. HAHN and F. STRASSMANN

[translated from 'Über den Nachweis und das Verhalten der bei der Bestrahlung des Urans mittels Neutronen entstehenden Erdalkalimetalle', *Naturwissenschaften* **27**, 11–15 (1939); from H.G. Graetzer, 'Discovery of Nuclear Fission', *Am. J. Phys.* **32**, 9–15 (1964)]

In a recent preliminary article in this journal[2] it was reported that when uranium is irradiated by neutrons, there are several new radioisotopes produced, other than the transuranic elements — from 93 to 96 — previously described by Meitner, Hahn, and Strassmann. These new radioactive products are apparently due to the decay of U^{239} by the successive emission of two alpha particles. By this process the element with a nuclear charge of 92 must decay to a nuclear charge of 88; that is, to radium. In the previously mentioned article a tentative decay scheme was proposed. The three isomeric radium isotopes with their approximate half-lives given, each decay to isomeric actinium isotopes, which in turn decay to thorium isotopes.

A rather unexpected observation was pointed out, namely that these radium isotopes, which are produced by alpha emission and which in turn decay to thorium, are obtained not only with fast but also with slow neutrons.

The evidence that these three new parent isomers are actually radium was that they can be separated together with barium salts, and that they have all the chemical reactions which are characteristic of the element barium. All the other known elements, from the transuranic ones down through uranium, protactinium, thorium, and actinium have different chemical properties than barium and are easily separated from it. The same thing holds true for the elements below radium, that is, bismuth, lead, polonium, and ekacesium (now called francium). Therefore, radium is the only possibility, if one eliminates barium itself.

In the following, the separation of the mixture of isotopes and the isolation of each species is described. From the changes in the activity of the various isotopes their half-lives can be found, and also the decay products can be determined. The half-lives of the daughter decay products cannot be fully described in this article however because of the complexity of the process. There are at least three, and probably four, isomeric decay chains, each one with three species. The half-lives of all the daughter products could not be thoroughly investigated so far.

Barium was of course always used as a carrier for the "radium isotope". As a first step, one can precipitate the barium as barium sulfate, which is the least soluble barium salt after the chromate. However, due to previous experience and some preliminary work, this method of separating the "radium isotope" by means of barium sulfate was not used. The reason was that this precipitate also carries with it a small amount of uranium, and also a not negligible quantity of actinium and thorium isotopes. These are the supposed decay products of the "radium isotope", and therefore they would prevent making a pure preparation of the primary decay products. Instead of the sulfate precipitate, barium chloride was chosen as precipitating agent, which is only very slightly soluble in strong hydrochloric acid. This method worked very well.

When uranium is bombarded with slow neutrons, it is not easy to understand from energy considerations how radium isotopes can be produced. Therefore, a very careful determination of the chemical properties of the new artificially made radioelements was necessary. Various analytic groups of elements were separated from a solution containing the irradiated uranium. Besides the large group of transuranic elements, some radioactivity was always found in the alkaline-earth group (barium carrier), the rare-earth group (lanthanum carrier), and also with elements in group IV of the periodic table (zirconium carrier). The barium precipitate was the first to be investigated more thoroughly, since it apparently contains the parent isotopes of the observed isomeric series. The goal was to show that the transuranic elements, and also U, Pa, Th, and Ac, could always be separated easily and completely from the activity which precipitates with barium.

1. For this reason, the irradiated uranium was treated with hydrogen sulfide, and the transuranic group was separated with platinum sulfide and dissolved in *aqua regia*. Barium chloride was precipitated from this solution with hydrochloric acid. From the remaining filtrate, the platinum was precipitated again with hydrogen sulfide. The barium chloride was inactive, but the platinum sulfide still had an activity of about 500 particles/minute. Similar experiments with the longer-lived transuranic elements gave the same result.

2. A precipitate with barium chloride was made using 10 g of unirradiated uranium nitrate. The U was in radioactive equilibrium with UX_1 + UX_2 (thorium and protactinium isotopes) and had an activity of about 400 000 particles/minute. The precipitate showed an activity of 14 particles/min; that is, once again, practically no activity. That means neither U, nor Pa, nor Th, comes down out of solution when the barium chloride crystallizes.

3. Finally, using a solution of actinium ($MsTh_2$) having an activity of about 2500 particles/min, a barium chloride precipitate was separated. This gave only about 3 particles/min which is also practically inactive.

In a similar way, the strongly active precipitates of barium chloride obtained from the irradiated uranium solution were carefully investigated. However, sulfide precipitates obtained from a neutral solution of the radioactive barium by means either of weak acetic acid or weak mineral acid [i.e. hydrochloric, nitric or sulfuric] were practically inactive, while lanthanum and zirconium precipitates had only slight activities whose origin could easily be traced to the activity of the barium precipitates.

A simple precipitate with $BaCl_2$ from a strong hydrochloric acid solution naturally does not allow one to distinguish between barium and radium. According to the reactions very briefly summarized above, the radioactivity which precipitates with the barium salts can only come from radium, if one eliminates barium itself for the time being as altogether too unlikely.

We now discuss briefly the graphs of the activities obtained with the barium chloride. They enable us to determine the number of "radium isotopes" present, and also their half-lives.

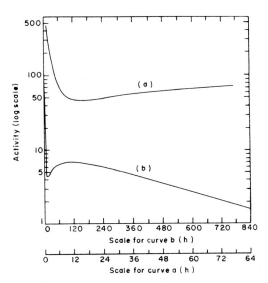

Fig. 1.   The three "Ra isotopes" after long irradiation. Curve a is the activity due to a 4-day irradiation, measured for about 70 h. Curve b is scaled down by a factor of 10 relative to the upper curve, showing the data for about 800 h.

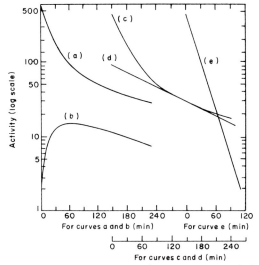

Fig. 2.   Determination of the half-life of Ra II (after short irradiation). Curve a shows the decay rate of all the Ra activity after 6-min irradiation. Curve b is the theoretical growth curve of the 2.5 h Ac II. Curve c is the difference $a - b$. Curve d is due to Ra III with a half-life of 86 min. Curve $e = c - d$, yielding a half-life of 14 min for Ra II.

Figure 1 shows the activity of the radioactive barium chloride after a 4-day irradiation of uranium. Curve *a* gives the measurements for the first 70 h; curve *b* gives the measurements on the same sample continued for 800 h. The lower curve is plotted on 1/10 the scale of the upper one. At first there is a rapid decrease of activity, which gradually levels off followed by a slow increase after about 12 h. After about 120 h, a very gradual exponential decrease of activity begins again, with a half-life of about 13 days. The shape of the curves shows clearly that there must be several radioactive substances present. However one cannot tell for sure what these may be; whether it is several "radium isotopes", or one "radium isotope" with a series of radioactive daughter products that determines the course of the activity.

The three isomeric "radium isotopes" which were previously reported in the earlier article were confirmed here. They are designated for the time being as Ra II, Ra III, and Ra IV (because of a presumed Ra I reported below). Their identification and the determination of their half-lives is explained briefly with the help of the figures. Figure 2 shows the radioactive decay of the "radium" after a 6-min irradiation of uranium. Curve *a* is the total activity, measured for 215 min. This curve is a composite of the activity from two "radium isotopes", Ra II and Ra III (compare Fig. 3), and also a small amount of actinium, which is formed by decay of Ra II. This latter substance, which is designated as Ac II, has a half-life of about 2½ h. This was shown in another experiment, which is not described here. The theoretical growth curve for such an actinium isotope resulting from Ra II is shown in the figure as curve *b*. Here the half-life of Ra II is taken to be 14 min, in anticipation of later results. When curve *b* is subtracted from curve *a*, then curve *c* in Fig. 2 is obtained. This remaining activity

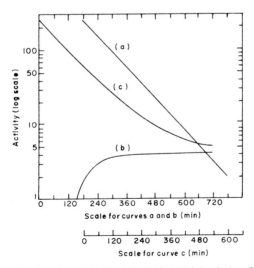

Fig. 3. Determination of the half-life of Ra III after 2½ h irradiation. Curve *a* shows the Ra III activity. Three hours after irradiation, actinium was removed. Curve *b* is the growth curve for long-lived Ac resulting from the decay of Ra III. Curve *c* = *a* − *b*, yielding a half-life of about 86 min for Ra III alone.

must now be due exclusively to the radium isotopes, mostly being due to the short-
lived Ra II, and with a slight contribution from Ra III with its longer half-life. The
latter has a half-life of about 86 min, as is seen in Fig. 3 later on. Curve *d* in Fig. 2
shows the activity due to Ra III. When *d* is subtracted from *c*, one finally obtains curve
e, which is the activity due to pure Ra II. It has an exponential decrease with a half-
life of 14 min. This value should be correct within ± 2 min.

Now we come to the identification and half-life determination of Ra III. A uranium
sample is irradiated for one hour or several hours. One finds a rapid decrease in activity
at first, then a still rather intense activity which decreases to one-half in about 100—
110 min, and then a further decrease. In order to show that this activity was also mostly due to a radium isotope, the following procedure was used. The "radium" was
separated from the irradiated uranium sample with barium chloride; after 2½ h, the
barium chloride was dissolved again, and reprecipitated. The short-lived Ra II has
completely decayed during this time, and the Ac II (2½ h half-life) which was formed
from Ra II in the barium chloride is removed in the recrystallization process. The
barium chloride still has considerable activity, so a "radium isotope" must still be
present. The procedure here is like that used by Meitner, Strassmann, and Hahn[3] for
the investigation of the artificial radioactive daughter products of thorium. The
resulting activity which remains is shown in Fig. 3, curve *a*.

During the first hours, the rate of decrease is almost exactly exponential, with a
half-life of about 86 min. A small residual activity remains, which is no doubt due to a
long-lived "actinium isotope" formed by the decay of Ra III. The supposed decay of
the actinium activity can be roughly estimated by the deviation of curve *a* from a
purely exponential one. This is shown in Fig. 3 as curve *b*. (It was also shown chemically that the decay of Ra III leads to an "actinium isotope" with a relatively long
life.) If one subtracts *b* from *a*, one obtains curve *c* for Ra III alone. It shows a very
nice exponential decrease with a half-life of 86 min. This value should be correct
within ± 6 min.

Now we come to the third "radium isotope", which is designated here as Ra IV.
In Fig. 1, the latter portion of curve *b* indicates a substance with a half-life of about 12
to 13 days. In a manner quite similar to that used for Ra III, it was shown that this
more slowly decreasing activity must be practically all due to a "radium isotope". A
lengthy irradiation of uranium was made, then the neutron source was removed, and
by waiting about one day the isotopes Ra II and Ra III were allowed to decay completely. If one makes a barium precipitate now, and as a precaution recrystallizes
again, then any activity found with the barium chloride can only be due to another
"radium isotope". Such an activity was always found, even after several days of waiting. The decay rate follows a characteristic pattern. It increases gradually for several
days, reaches a maximum, and then decreases with a half-life of about 300 h (12.5
days).

In Fig. 4 are shown several such curves. The sample for curve *c* was prepared from
a uranium solution which had been irradiated with low intensity, and the other curves
are due to barium precipitates from more intensely irradiated uranium solutions. (The
curves cannot be used to determine the relative intensity factor directly, since the geometrical arrangement was not identical. Under identical conditions, such as equal
amounts of uranium being irradiated, etc., we found that the relative intensity factor
was about 7.) The shapes of the three curves are very similar. The growth of activity

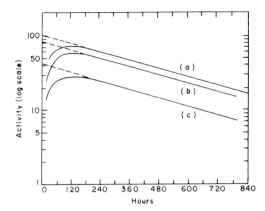

Fig. 4. Determination of the half-life of Ra IV by differing times and intensities of irradiation. Curve a, 4-day high intensity irradiation. Curve b, 2.6-day high intensity irradiation, with chemical separation of Ra 15 h after irradiation was completed. Curve c, 2.6-day low intensity irradiation. For curves a, b, and c, respectively, the half-lives are about 311, 310, and 300 h.

has a half-life of less than 40 h, and the decay about 300 h. However, the long-lived "Ra IV" doubtlessly has a half-life of somewhat less than 300 h, because the Ac IV which is mostly responsible for the initial growth of activity probably decays to a long-lived "thorium isotope". Therefore the half-life of Ra IV cannot be determined precisely, but a value of 250–300 h is probably close to being correct. From curves a, b, and c, one can see clearly that the beta rays of Ra IV are much less penetrating than those from its daughter product, since otherwise such a sharp increase would not occur.

To summarize our results, we have identified three alkaline earth metals which are designated as Ra II, Ra III, and Ra IV. Their half-lives are 14 ± 2 min, 86 ± 6 min, and 250–300 h. It should be noted that the 14-min activity was not designated as Ra I nor the other isomers as Ra II and Ra III. The reason is that we believe there is an even more unstable "Ra" isotope, although it has not been possible to observe it so far. In our first article about these new radioactive decay products we reported an actinium isotope with a half-life of about 40 min. Our initial assumption was that this least stable actinium isotope had resulted from the decay of the least stable radium isotope. In the meantime, we have determined that the 14-min radium (previously given as 25 min) decays to actinium with a 2.5-h half-life (previously given as 4 h). However, the less stable actinium isotope mentioned above is also present. Its half-life is somewhat shorter than previously reported, perhaps a little less than 30 min. This "actinium isotope" cannot result from the decay of the 14-min, 86-min, or the long-lived "Ra". Also this "actinium isotope" can be shown to be present after only a 5-min irradiation of uranium. The simplest explanation is to assume the formation of a "radium isotope" whose half-life must be shorter than 1 min. If it had a half-life longer than 1 min, we should have been able to detect it. We searched for it very carefully. Therefore we designate this heretofore unknown parent of the least stable "actinium isotope" as "Ra I". With a more intense neutron source it should no doubt be detectable.

The decay scheme which was given in our previous article must now be corrected. The following scheme takes into account the needed changes, and also gives the more accurately determined half-lives for the parent of each series:

$$\text{"Ra I"?} \xrightarrow[< 1 \text{ min}]{\beta} \qquad \text{Ac I} \xrightarrow[< 30 \text{ min}]{\beta} \text{Th?}$$

$$\text{"Ra II"} \xrightarrow[14 \pm 2 \text{ min}]{\beta} \qquad \text{Ac II} \xrightarrow[2.5 \text{ h}]{\beta} \text{Th?}$$

$$\text{"Ra III"} \xrightarrow[86 \pm 6 \text{ min}]{\beta} \qquad \text{Ac III} \xrightarrow[\text{several days?}]{\beta} \text{Th?}$$

$$\text{"Ra IV"} \xrightarrow[250-300 \text{ h}]{\beta} \qquad \text{Ac IV} \xrightarrow[< 40 \text{ h}]{\beta} \text{Th?}$$

The large group of transuranic elements so far bears no known relation to these [isomeric] series.

The four decay series listed above can be regarded as doubtlessly correct in their *genetic* relationship. We have already been able to verify some of the "thorium" end products of the isomeric series. However, since the half-lives have not been determined with any accuracy yet, we have decided to refrain altogether from reporting them at the present time.

Now we still have to discuss some newer experiments, which we publish rather hesitantly due to their peculiar results. We wanted to identify beyond any doubt the chemical properties of the parent members of the radioactive series which were separated with the barium and which have been designated as "radium isotopes". Therefore we have carried out fractional crystallizations and fractional precipitations with the active barium salts using a procedure which is well-known for concentrating (or diluting) radium in barium salt solutions.

Barium bromide increases the radium concentration greatly in a fractional crystallization process and barium chromate even more so when the crystals are allowed to form slowly. Barium chloride increases the concentration less than the bromide, and barium carbonate decreases it slightly. When we made corresponding tests with radioactive barium samples which were free of any later decay products, *the results were always negative. The activity was distributed evenly among all the barium fractions*, at least to the extent that we could determine it within an appreciable experimental error. Next a pair of fractionation experiments were done, using the radium isotope ThX and also the radium isotope $MsTh_1$. These results were exactly as expected from all previous experience with radium. Next the "indicator (i.e., tracer) method" was applied to a mixture of purified long-lived "Ra IV" and pure $MsTh_1$; this mixture with barium bromide as a carrier was subjected to fractional crystallization. *The concentration of* $MsTh_1$ *was increased, and the concentration of* "Ra IV" *was not*, but rather its activity remained the same for fractions having an equivalent barium content. We come to the conclusion that our "radium isotopes" have the properties of barium. As chemists we should actually state that the new products are not radium, but rather barium itself. Other elements besides radium or barium are out of the question.

Finally we have made a tracer experiment with our pure separated "Ac II" (half-life about 2.5 h) and the pure actinium isotope MsTh$_2$. If our "Ra isotopes" are not radium, then the "Ac isotopes" are not actinium either, but rather should be lanthanum. Using the technique of Curie[4], we carried out a fractionation of lanthanum oxalate, which contained both of the active substances, in a nitric acid solution. Just as Mme. Curie reported, the MsTh$_2$ became greatly concentrated in the end fractions. With our "Ac II" there was no observable increase in concentration at the end. We agree with the findings of Curie and Savitch[5] for their substance with a 3.5 h half-life (although not uniformly so) that the product resulting from the beta decay of our radioactive alkaline earth metal is not actinium. We want to make a more careful experimental test of the statement made by Curie and Savitch that they increased the concentration in lanthanum (which would argue against an identity with lanthanum) since in the mixture with which they were working there may have been a false indication of enrichment.

It has not been shown yet if the end product of the "Ac-La sample", which was designated as "thorium" in our isomeric series, will turn out to be cerium.

The "transuranic group" of elements are chemically related but not identical to their lower homologs, rhenium, osmium, iridium, and platinum. Experiments have not been made yet to see if they might be chemically identical with the even lower homologs, technetium, ruthenium, rhodium, and palladium. After all one could not even consider this as a possibility earlier. The sum of the mass numbers of barium + technetium, 138 + 101, gives 239!

As chemists we really ought to revise the decay scheme given above and insert the symbols Ba, La, Ce, in place of Ra, Ac, Th. However as "nuclear chemists", working very close to the field of physics, we cannot bring ourselves yet to take such a drastic step which goes against all previous experience in nuclear physics. There could perhaps be a series of unusual coincidences which has given us false indications.

It is intended to carry out further tracer experiments with the new radioactive decay products. In particular a combined fractionation will be attempted, using the radium isotope resulting from fast neutron irradiation of thorium (investigated by Meitner, Strassmann, and Hahn, Ref. 3) together with our alkaline earth metals resulting from uranium. At places where strong neutron sources are available, this project could actually be carried out much more easily.

Acknowledgments. In conclusion we would like to thank Miss Cl. Lieber and Miss I. Bohne for their efficient help in the numerous precipitations and measurements.

NOTES

[1] From the Kaiser Wilhelm Institute for Chemistry, at Berlin-Dahlem. Received 22 December 1938. Translation published with the permission of the editor of *Naturwissenschaften* and of the authors.

[2] O. Hahn and F. Strassmann, *Naturwiss.* **26**, 756 (1938).

[3] L. Meitner, F. Strassmann, and O. Hahn, *Z. Physik* **109**, 538 (1938).

[4] Mme. Pierre Curie, *J. Chim. Phys.* **27**, 1 (1930).

[5] I. Curie and P. Savitch, *Compt. Rend.* **206**, 1643 (1938).

2

From antiquity to Paracelsus

INTRODUCTION

Transmutation can be defined as the conversion of one element or nuclide into another whether naturally as in radioactive disintegration or artificially by nuclear bombardment. But transmutation may also be defined as the allegedly possible conversion of base metals into gold and silver by alchemy. That there are differences between these two notions may be obvious; not so, however, the similarities beyond the term 'transmutation'. Before deciding whether there are tenable connections between ancient and modern transmutation, it will be necessary to explore what meaning this had in antiquity. And this brings us to alchemy.

Alchemy may be considered the art of transmutation or perfection of metals described in terminology at once physical and mystical. These components can be distinguished, but not entirely separated one from another. The language and symbolism used by the alchemists is quite unintelligible in any physical sense. While this style and approach may have been adopted in part to protect the esoteric secrets of the cult from being profaned, it also reflects the character of alchemistic thought. The cosmos was construed as a unity without any essential tension between the physical and the mystical–religious domain. The possibility of transmutation was linked to this alchemical *Weltanschauung* presupposing a fundamental unity of all things and a common material principle which remained constant throughout change. Moral qualities 'imperfect' and 'base' were associated with metals as with man. Perfection was the natural final state of all beings, a goal envisioned as more readily attainable through the application of the so-called philosophers' stone.*

ORIGIN OF ALCHEMY

It is more likely that alchemy arose independently within several cultures than that it had a unique historical origin. Whatever its actual genesis there can be little doubt that it flourished and developed within the Hellenistic culture of Alexandria in Egypt for nearly a millennium from about 300 BC. It remains undecided whether the name 'alchemy' is derived from the Greek Χημεια through the Arabic *khimiya* as referring to those metallurgical and kindred practices in vogue in Greco-Roman Egypt or if it once designated *Khem*, the ancient name for Egypt in reference to the black soil characteristic of that country. Either version, however, leads to Egyptian fosterage, a view consistent with tradition and with the general acknowledgement of Hermes Trismegistos, representing Thoth, the Egyptian god of learning, as the father of this 'art'. Alexandria was a cosmopolitan center like London, Paris, Vienna or New York in the modern era. It was here that abstract Greek philosophy could intermingle with Egyptian technology and religious mysticism from the Orient, producing that amalgam upon which alchemy thrived.

GREEK PHILOSOPHY

One of the most characteristic features of Greek philosophy was the four-element theory of earth, air, fire and water, with the addition of a fifth extraterrestrial element included when treating the entire cosmos. More fundamental than the elements, however, was a substratum or common material principle which became known traditionally as 'prime matter'. The predominance of the element 'earth', for example, would tend to make a body solid; that of 'water' would tend to yield liquids, and so forth. Every body was understood as consisting of a mixture of these elements. If the proportion among the types of elements constituting a given body were to shift, say, from a predominance of the element 'earth' to one favoring 'water', the body would accordingly have undergone a transformation. Such a change would be achieved in virtue of the dual qualities assumed to be characteristic of each element (see Fig. 1). Thus water could be transmuted into air as an invisible vapor by replacing its characteristic quality 'cold' by boiling it until it becomes 'hot'. We would call it steam.

EGYPTIAN ARTISANS

Recent popular interest in Egyptology has made it abundantly clear that

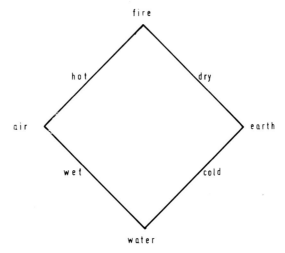

Figure 1. The four elements.

skilled artisans were working with precious metals over three millennia ago. These metallurgical workers and goldsmiths prepared silver and golden ornaments and gilded objects for the ruling class and the rich, while producing substitutes for others. The first alchemists were probably just such artisans attempting to make artificial gold and silver for the general public. That such inferior products should in principle be possible was consistent with the four-element theory and with the scale of perfection associated with religious mysticism.

RELIGIOUS MYSTICISM

The ancient tradition perceiving a unity between the astrological macrocosm and the terrestrial microcosm has roots also in that time and culture. Astrological influences were linked not only to human activity and destiny but also to the alchemical replacement of one metallic state for another. This essential unity throughout the cosmos, implying a correspondence between what is spiritual and what is physical, became a central feature of alchemistic thought. It was accordingly the proper destiny of any microcosmic being to achieve that state of perfection characteristic of the macrocosm. The perfection of man was sought either in the spiritual realm or in extending physiological life indefinitely through potions such as the elixir of life. Gold was the only metal that had reached the perfect metallic state; and the other metals which had not as yet reached this, their proper destiny, were to that extent deficient or 'base'. The artisans as alchemists, so it was thought, could perhaps hasten this process of perfection. In searching for the philosophers' stone they were really pursuing a principle or medium for perfectibility.

TRANSMUTATION

There is perhaps no concept more characteristic of the alchemical tradition than transmutation. This can be considered from its physical and practical aspect in spite of the inherently mystical component of alchemy. The transmutation of metals may well have seemed commonplace to the alchemists, a case in point being the deposition of copper on iron when immersed in a solution of a copper salt such as blue vitriol. Yet the two metals merely exchange places! The existence of white and yellow alloys of copper with arsenic and other substances may well have suggested that copper might be transmutable into silver or gold. That gold should have been considered the most perfect metal with silver just one step removed in perfection is not difficult to understand on the basis of their natural beauty alone. Gold does not tarnish and it resists the action of fire, most corrosive liquids and even sulfur.

Following the accepted theory of transmutation by means of qualitative change, the alchemists made use of visual indicators to trace the progress of induced change. For successful transmutation, the sequence of ever increasing perfection should be followed; a chromatic hierarchy from black through a stage of whitening to a yellowing followed by the final process yielding violet, purple or red as the highest color. For example, copper could be converted to the black oxide and then whitened with mercury or arsenic to form a silvery alloy. This color sequence became a standard method of tracing the degree of perfection throughout the transmutation sequence. The preparation of the brilliant red sulfide of mercury (HgS), known in nature as cinnabar, held particular fascination for the alchemists, since sulfur and mercury were considered the basis of all metals.

SYMBOLISM

Most of the metals associated with alchemy — gold, silver, lead, copper, tin and iron — were known long before the rise of alchemy. The liquid metal, mercury, was familiar to the ancient Chinese and has been found in Egyptian tombs older than three millennia. Sulfur too was known to antiquity. As a further indication of the cosmic unity postulated by the

Metals.	Planets, &c.¹³	Symbols.
Gold	Sun	☉
Silver	Moon	☽
Mercury	Mercury	☿
Copper	Venus	♀
Iron	Mars	♂
Tin	Jupiter	♃
Lead	Saturn	♄

Figure 2. Symbols of the metals, from H. S. Redgrove, *Alchemy, Ancient and Modern*, p. 27. (Courtesy University Books Inc.)

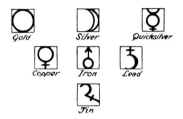

Figure 3. Symbols of the metals, from J. Read, *Prelude to Chemistry*, p. 182. (Courtesy Bell & Hyman Ltd.)

alchemical philosophy, these metals were symbolically associated with the celestial bodies (see Figs. 2 and 3). The Sun and Moon were linked with the two noblest metals, gold and silver, while Saturn referred to lead. The other planets, Venus, Jupiter, Mars and Mercury, were generally associated with copper, tin, iron and metallic mercury, respectively.

> In those days the metals were suns and moons . . . Gold was Apollo, sun of the lofty dome; silver, Diana, the fair moon of his unresting career . . .; quicksilver was the wing-footed Mercury, Herald of the Gods . . .; iron was the ruddy-eyed Mars, in panoply complete; lead was heavy-lidded Saturn, quiet as a stone . . . and so forth in not un-meaning mystery. [Reference 2.]

The unity of matter in alchemical thought is reinforced by the symbol of the tail-devouring serpent, Ouroboros (see Fig. 4) expressing 'The One is The All' (ἐν τὸ πᾶν). The correspondence between the microcosm and the macrocosm is represented here, as is the principle of *materia prima* adopted from Greek philosophy.

Figure 4. Ouroboros, the tail-eating serpent – a key symbol of the alchemical tradition. (From H. J. Sheppard, *Ambix*, **10**, 85 (1962); courtesy Society for the History of Alchemy and Chemistry.)

CHINESE ALCHEMY

Alchemy can be traced back not only to Hellenistic Egypt but to other cultures as well, although it is doubtful that gold making and the transmutation of metals played such a significant role elsewhere. Indian alchemy was characterized by the preparation of medicinal elixirs for healing and curing certain diseases. Their religious convictions precluded the necessity of similar elixirs for achieving immortality. Perhaps the most important alternative alchemy came from China – like Alexandria, dating back to about 300 BC. Although the transmutation of base metals into gold is mentioned, it is medicinal alchemy that predominates. Indeed, the simple fact that Chinese alchemy was associated with medicine rather than metallurgy would seem to confer priority upon it, if this were considered a question of interest. Their *leit-motiv* was a belief in the possibility of attaining physiological immortality through the medium of what became known as the 'elixir of life'. There seems to have been an affinity between Chinese alchemy and that mystical religion which degenerately developed from the Taoist philosophy about the fifth century BC. Here again we meet a belief in the order, unity and uniformity throughout the cosmos. There is evidence that the Chinese attempted metallic transmutation as early as the third century BC and this continued to develop as a secondary aspect of their alchemy. The central feature of Chinese alchemy was the introduction of quasi-medicinal substances, generally derivatives or compounds of mercury, which were to act as catalysts either for bringing about metallic transmutation or, if ingested in liquid form as potable gold, for prolonging life physiologically and allegedly even bestowing immortality in some cases. Most likely those who imbibed these potions were poisoned fatally.

Perhaps the best known of the Chinese alchemists was Ko Hung who flourished about AD 320. His recipes for elixirs were mostly based upon compounds of arsenic or mercury, and he clearly distinguished the preparation of potable gold – the elixir of life – from the production of artificial cinnabar to be used for the transmutation of base metals into silver and gold.

In general, then, the Chinese alchemists were quite familiar with mercury, sulfur, and combinations of both, a favored type of elixir being that based upon cinnabar. Although they did deal with the transmutation of metals, their main preoccupation was consistently with physical immortality in contrast to the mystical–spiritual perfection associated with western religious philosophy.

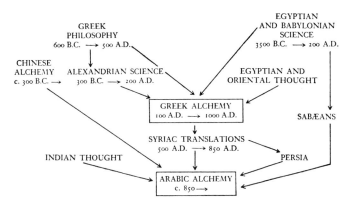

Figure 5. Table of alchemical tradition. (From F. S. Taylor, *The Alchemists*, p. 67; courtesy Henry Schuman Inc.)

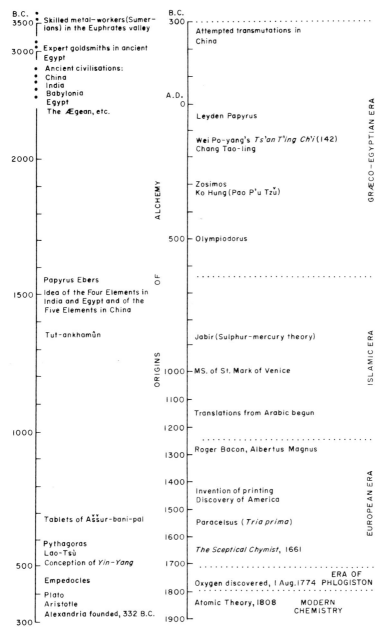

Figure 6. Time chart of alchemy. (From J. Read, *Prelude to Chemistry*, p. 35; courtesy The Macmillan Co., Inc.)

ALCHEMICAL WRITINGS (see Fig. 6)

Most of the records of Chinese alchemy are included in the Taoist canoni-
cal writings. Perhaps the most famous alchemical book in the Chinese
tradition is the *Great Secrets of Alchemy* from the early seventh century
AD concerning the production of elixirs for the attainment of immortality. Two of the earliest and best known works from Hellenistic alchemy
are probably *Physica et Mystica* by Pseudo-Democritus (Bolos of Mende)
written about 200 BC and containing recipes for making gold and silver,
as well as the encyclopaedic works by Zosimus of Panopolis, another
Egyptian, written about five centuries later. The latter must have been
roughly contemporary to his Chinese counterpart Ko Hung, and both
exhibit about the same degree of knowledge of chemical operations.
Zosimus provided elaborate descriptions of how base metals were to be
ennobled to gold by a process of death and resurrection. The transmutation procedure, in order to be successful, was supposed to pass through
the previously mentioned color sequence – black, white, yellow and
purple. The resultant tinted imitation was considered by Zosimus to be
even superior to native gold or silver both in color and in purity. The later
alchemists of the Hellenistic tradition also began to concern themselves,
as the Chinese had all along, with a substance which could catalytically
influence the process of transmutation. Zosimus called such a substance
the 'tincture', but it was variously referred to as the 'powder' or the
'elixir' and became known ultimately as the 'philosophers' stone'.

ARABIC ALCHEMY

One of the most influential of alchemical writings was the *Emerald
Tablet*, attributed to Hermes the Greatest (see Fig. 7). Although Arabic
alchemy derived partly from Greek roots, it was not Pseudo-Democritus
but rather the even more remote Hermes Trismegistos who was acclaimed
by the Arabic alchemists to be the fountainhead of their art. Yet he
figures only minimally in Greek alchemical writings. Such attribution
reflects the tendency of early writers, and not only those concerned with
alchemy, to ascribe their works to some very dignified or historical figure
so as to add authority and stature to what they have presented. Clearly
someone wrote the *Emerald Tablet*, part of the *Book of the Secret of
Creation*, but Hermes, as has been pointed out, is very probably the
personification of an Egyptian deity. That we still speak of hermetically
sealed flasks and the like reflects the tradition attributing to this figure
even the invention of a magic seal to keep vessels airtight (Reference 3).

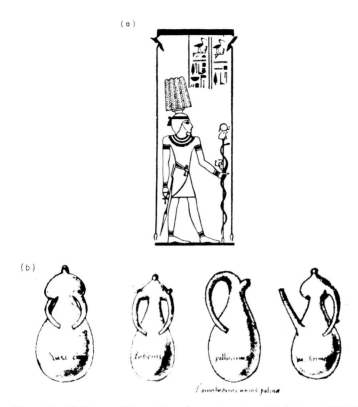

Figure 7. (a) Hermes Trismegistos, from a temple at Pselcis, *ca.* 250 BC.
The symbolism is noteworthy: the figure holds the *'ankh* or *crux ansata* in
one hand, and in the other a staff with a serpent, scorpion, hawk's head,
and asp enclosing a circle. (From J. Read, *Prelude to Chemistry*, p. 34;
courtesy Bell & Hyman Ltd.)
(b) Hermetic vessels, from a sixteenth-century manuscript. (From F. S.
Taylor, *The Alchemists*, facing p. 119; courtesy Henry Schuman Inc.)

Arabian alchemy dates from about the eighth century AD and inherited
aspects from both the Chinese and the Alexandrian. It was probably the
Mongols and the trade routes which provided the Chinese connection. The
Hellenic linkage is complex but better documented. Greek manuscripts
including many alchemical works were translated first into Syriac by the
Nestorian Christians after AD 500 and again into Arabic after AD 750, in
centers of activity such as Baghdad and Harran, following the rise of
Islam.

The most famous of these was no doubt Jabir ibn Hayyan (Geber, fl. eighth century) whose name was adopted as a figurehead for a very active Bourbaki-type sect of Arabian alchemists from the ninth century. Probably the greatest of the Arabic alchemists was the Persian physician Al Razi (Rhazes, fl. 900) of Baghdad.

ARABIAN ALCHEMICAL PHILOSOPHY

Arabian alchemy was even more concerned with the manufacture of gold *per se* than either of its predecessors had been. The Arabs adopted the Chinese notion of a catalytic elixir and made serious attempts to produce gold by means of this agent. This quasi-medicinal medium was to be added to base metals to assist in the perfection of metals, but if ingested in liquid form it could also act as the elixir of life. The school of thought associated with Geber, however, maintained that the elixir was not exclusively of a mineral nature but could contain animal or even vegetable components.

The four-element theory and the substrate of prime matter were adopted from the Hellenistic tradition but this too underwent some modification. Sulfur and mercury were postulated as intermediate stages arising from the change of earth into fire (yielding the sulfur exhalation) and of water into air (yielding the mercury exhalation). These two constituents, in differing proportions and degrees of purity, representing the basic principles of combustibility and fluidity respectively, allegedly gave rise in turn to the various metals. Gold itself was supposed to consist of the contrary elements fire and water in the sense that it combined within itself both sulfur, the principle of combustibility, and mercury, the principle of liquidity.

LATIN ALCHEMY

As part of the reawakening of western Europe in the second millennium, Christian scholars of the twelfth century began to translate the Arabic and Greek texts including the literature of alchemy. It is probably during this period that the name 'alchemy' began to be used for what had previously been known as the 'art' of transmutation. The works of Geber, Rhazes, and other Arabs became available to scholars by the middle of the twelfth century. Before the dawn of the fourteenth century, alchemy was being commented upon and discussed throughout Europe, as is exemplified by Roger Bacon in England, Albert the Great in Germany, Arnold of Villanova in France, and Raymond Lully in Spain.

These were in the first instance transmitters of alchemical tradition rather than true alchemists. Roger Bacon was a firm supporter of the sulfur—mercury theory of metals, and he repudiated all attempts to adulterate mineral substances with either animal or vegetable components in the production of the philosophers' stone. Albert the Great included alchemical accounts in his systematic works and considered alchemy to be a general science of potential utility. His information is largely derived from the influential *De Anima* of Pseudo-Avicenna. Albert recognized a celestial influence upon chemical transformations, and mentioned the artificial production of red vermilion or cinnabar by subliming mercury, sulfur and sal-ammoniacum. He doubtlessly believed in the possibility of metallic transmutation, but considered it very difficult to realize. It is attributed to Arnold of Villanova that all metals can be reduced to mercury, construed as the first matter of all metals; and hence transmutation is quite feasible. Lully is also supposed to have maintained a reductionist viewpoint, noting that the transmutation of metals can be achieved only after reducing the minerals to their first matter. The secret of the philosophers' stone, he thought, lay in the extraction of the mercury within silver and gold.

Perhaps the most influential alchemist of this period was Pseudo-Geber (fl. 1310) who wrote practical recipes for transmutation and whose works were adopted by alchemists in the manner of a textbook. In a commentary entitled *The New Pearl of Great Price*, attributed to one Petrus Bonus of Italy and supposed to have been written about 1330, Peter follows Pseudo-Geber in observing that all metals consist of mercury and sulfur. But if there is only one mercury, he claims that there are two kinds of sulfur. The object of alchemy then is not to reduce imperfect metals to their first matter but to assist nature in removing the impure sulfur and in thereby bringing them to perfection. Only gold has already achieved this highest state. Bonus wisely remarks that it was both practical and useful that there were in fact such imperfect metals, since without copper, iron, tin, lead, bronze and so forth neither tools nor utensils could be fashioned.

The Midas legend suffices to draw our attention to what a sorry state of affairs there would be if all metals were ever perfected in this way. Nevertheless, the alchemist's position is simply that such a state should be possible at least in theory, and that it was the role of the philosophers' stone to assist nature in realizing this allegedly ideal and final state.

LATER ALCHEMY

By the middle of the fourteenth century, the number of works dealing

Plate 4. 'Der Alchymist', from the engraving by Edm. Hellmer.

with alchemy increased greatly and the dominant thrust had become the manufacture of gold. For the next few centuries practical attention was also given to the preparation of mineral acids in alchemical fashion. The mystical side of alchemy, though, tended to become increasingly obscure, and by the fifteenth century was very much a self-contained esoteric cult. Symbolism and allegory became increasingly complex, and writings tended to be repetitive of what had preceeded, so that nothing really new emerged.

3

From Paracelsus to late nineteenth century

INTRODUCTION

Tracing the history of alchemy in its relation to the question of transmutation has made it clear that this 'art' is embedded in a complex matrix
of ideas which cannot really be separated. There are aspects both physical
and mystical. By the fifteenth century, the idea of immortality and the
elixir of life, derived from the orient, had become blended with their
occidental counterparts of perfectibility and the philosophers' stone as
the proper medium for transmutation. Other aspects include the macrocosm—microcosm analogy whereby celestial events can influence terrestrial processes, including those of the alchemists. The very possibility of
transmutation, of course, hinges upon there being some material substrate
underlying the change, and indeed the unity of matter forms another basic
theme in conjunction with some first-matter principle. Alchemy was also
associated with gold artisans and those who may have construed the 'art'
as merely a get-rich gimmick and even misused it to further their own
ends.

From this amalgamation of ideas several key notions, albeit interlocked,
can be identified. These at once represent the essence of alchemical transmutation and can fruitfully be traced up to the modern period. Seen from
the perspective of its history and philosophy, alchemical transmutation
combines both the unity of matter and the existence of some transmuting
medium as two essential principles. The first is traditionally associated
with *materia prima* and the second with the philosophers' stone. We can
therefore identify three essential features which will serve as guidelines
throughout. The first is the notion of transmutation as such, which can be
construed either alchemically or in terms of sub-atomic change.* The
second is the medium by which to achieve transmutation. This is traditionally understood as a substance but can be interpreted functionally.

Indeed, as we shall see, it was often the *method* which proved to be the effective 'medium' for identifying natural transmutation and for stimulating artificial transmutation by bombardment. Finally, there is the notion of unity of matter associated with some first-matter principle. Here too, this can be understood in terms of the elements of the ancients or as the structural principles of the chemical elements of modern science. These three features, then, may be seen as common to both the older and the newer alchemy. Let us now trace the relevant developments over the intervening centuries. While something of traditional alchemy was carried along during this period, it is the branching effect into medicine, pharmacy and chemistry which is perhaps most characteristic.

IATROCHEMISTRY

Traditional alchemy of the fifteenth century had degenerated largely into either a gold-making cult or one of symbolic mysticism. Yet it had not always been so. Alchemy had been practised in China primarily for purposes more closely related to medicine than to the transmutation of metals. This is not to say that early western alchemists lacked any interest in a possible medicinal role for their art. Roger Bacon had already linked alchemy with the search for the prolongation of life, and Arnold of Villanova had medical overtones in his alchemical writings. In light of this, the sixteenth century interest in the practical utilization of alchemy for medicinal purposes — the manufacture of curative chemical products — can be seen as an occidental revival and extension of this characteristically oriental concern. Iatrochemistry or medico-chemistry represents a tradition of the sixteenth and seventeenth centuries which maintained that the proper goal of alchemy was the production of chemical medicines. The founder and figurehead of this tradition was Paracelsus.

PARACELSUS

The Swiss medical doctor, Theophrastus von Hohenheim (1493–1541), otherwise known as Paracelsus, did take alchemy seriously, but as a system of thought involving the fundamental unity of the microcosm with the macrocosm. The art of transmuting metals and the search for chemical medicines were but two aspects of this larger picture, and he clearly emphasized the latter. He emerged on the scene at the age of 33 and died before reaching 50, but his voluminous writings attest to the greatness of this polymath. It was not his goal to free chemistry from the bonds of alchemy but to provide new directions. As he saw it, the true function of

Plate 5. Paracelsus (Theophrastus von Hohenheim), from a woodcut by Hirschvogel *ca.* 1540. (Courtesy Deutsches Museum, Munich, from the original in the Albertina, Vienna.)

alchemy was to prepare those principles which can restore harmony between man and the cosmos. He thus represents the beginning of the parting of the ways dividing those who tended to pursue chemistry and produce useful medicines from those who followed alchemy for metallic transmutation. Paracelsus and his followers made it their goal to manufacture chemical medicines, including potable metallic ones, to cure diseases which traditional Galenic medicine could not cope with. Although they continued to work within the language of alchemy, their introduction of chemical remedies led directly to modern pharmacology.

Their concern for chemical remedies reflects a rejection of the Galenic medical tradition current at the time. This was based upon the Hippocratean theory of the 'four humors' as given by Galen (fl. AD 175). Disease was thought to be the result of an imbalance among these four fluids normally in balance: blood, phlegm, yellow bile and black bile. Hence a cure was to be achieved by restoring the internal balance. Remedies were traditionally of plant extraction and more general than specific. In contrast to this tradition, the balance which concerned Paracelsus was not simply an intrasomatic one but that between man and his cosmic environment. Diseases were construed by Paracelsus as being caused by external agencies and directly related to bodily malfunctions, thought to be essentially chemical in kind. It was these specific disease entities and these specifically chemical malfunctions which Paracelsus sought to neutralize and remedy by specific preparations whether metallochemical or phytochemical in origin.

THE *TRIA PRIMA*

During the ninth century passage of alchemy into Islam, it will be recalled, the four-element theory underwent a considerable modification. The opposed elements fire and water had assumed a new guise in the sulfur–mercury theory of metals in which sulfur was the principle of combustion and mercury was that of metallicity. As a corollary, matter was clearly distinguished from its properties. Seven centuries later, Paracelsus made some important adjustments to this theory. Each of the four elements had been associated with a pair of qualities, but he accepted only one single property as characteristic of an element. Fire, for example, is not both hot and dry but just hot. More important, however, was his addition of a third hypostatical principle, salt (or magnesia), to the theory of Geber. Furthermore, this extended sulfur–mercury–salt theory was applicable not only to metals but could be used for all substances. Thus while sulfur remained the principle of combustibility or the inflammable state, mercury became the principle of vapor and liquidity in general or the vaporous and fluid

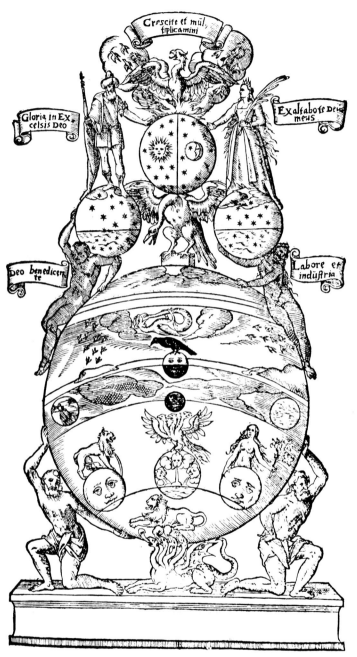

Figure 8. The Philosophers' Stone, frontispiece from A. Libavius, *Alchymia* (1606). (Courtesy Deutsches Museum, Munich.)

state. The third member of this *tria prima*, salt, was the principle of solidity or the state of fixation. The four-element theory, thus modified with the *tria prima*, was one of the major contributions of Paracelsus and iatrochemistry. These three principles were regarded as constituting all things and determining the state in which matter can occur; hence a healthy body differed from an unhealthy one in having them in a proper balance. This balance in turn is conditioned by influences outside the individual body, the astral connection, including the stars.

THE CHEMICAL ALCHEMISTS

Many of those who were associated with Paracelsus and iatrochemistry stood at the turning point between alchemy and chemistry. They developed chemical methods beyond distillation alone and used such techniques as the color test and the flame test. Only a few of these can be treated individually.

A. Libavius (fl. 1580) accepted the iatrochemical doctrine but was a firm believer in the transmutation of metals. He directed most of his attention, however, to the preparation of better medicines. His celebrated *Alchymia*, first published in 1595, contains a full account of chemical knowledge of his day and served as an authoritative and comprehensive textbook.

Basil Valentine (J. Thölde) was a rather typical Paracelsian emphasizing the macrocosm–microcosm analogy and the significance of chemical medicines. His popular and influential *Triumphal Chariot of Antimony* (1604) strongly reflects a Paracelsian influence. Thölde also described the precipitation of copper from solution by iron as an example of natural transmutation. He was well aware of the method used to prepare mercury by distillation of a mixture of sublimate and caustic lime. Quite apart from any alchemical concerns, mercury was used technically for the extraction of gold and silver by the 'amalgamation' process and for gilding.

J. B. van Helmont (fl. 1620) had studied Galenic medicine but later joined the iatrochemists, although he did not accept the sulfur–mercury–salt theory of Paracelsus. Fire he dismissed as an element or as anything material at all, but water was his *materia prima*. Helmont also had an affinity for the mystical, and he made claims to having been adept at alchemy.

J. R. Glauber (fl. 1640) also accepted the main iatrochemical doctrine but firmly believed that gold could be produced by the methods of alchemy. He made many contributions to the incipient science of chemistry, discovering new substances and their properties and inventing laboratory processes and equipment. Probably Helmont and Glauber, more than any others, are responsible for laying the foundations for a true chemical philosophy.

TRANSMUTATIONS OF THE ALCHEMISTS

With the dawn of iatrochemistry the attention of many alchemists was partly diverted from transmutation and the making of gold, but not completely. 'Main line' alchemy carried on too. There were many cases of apparent transmutation, but what they had produced were mainly alloys that resembled the noble metals. The alchemists also made many attempts to find the philosophers' stone. The nature of this long-sought substance seems to have been generally allied to mercuric sulfide (HgS), the only important ore of mercury. As cinnabar or vermilion this is reddish and takes either a hexagonal crystalline form or that of a powder, but as metacinnabarite it forms either black cubical crystals or an amorphous powder.

No less an authority than Helmont in the first half of the seventeenth century professed to have the power of the philosophers' stone and claimed to have successfully transmuted metals on occasion. Of particular interest is his account of the transmutation of mercury into gold.

[The Stone of the Philosophers] . . . was of colour such as in Saffron in its Powder, yet weighty, and shining like unto powdered Glass: There was once given unto me one fourth part of one Grain: But I call a Grain the six hundredth part of one ounce: This quarter of one Grain therefore, being rouled up in Paper, I projected upon eight Ounces of Quick-silver made hot in a Crucible; and straightway all the Quick-silver, with a certain degree of Noise, stood still from flowing, and being congealed, setled like unto a yellow Lump: but after pouring it out, the Bellows blowing, there were found eight Ounces, and a little less than eleven Grains of the purest Gold. [Reference 4.]

Another well-known case of transmutation is that described and reported in 1667 by the court physician Helvetius.

. . . there was a hissing sound and a slight effervescence, and . . . the whole mass of lead had been turned into the finest gold. Before this transmutation took place, the compound became intensely green . . . When it cooled it glittered and shone like gold. [Reference 5.]

THE MYSTICAL ALCHEMISTS

There were also alchemists who tended to adopt more of a mystical approach to the entire enterprise. The great mystic J. Boehme (fl. 1610),

Figure 9. Macrocosm and microcosm, plate from R. Fludd, *Metaphysica*
 . . . *Historia* (1617). (Courtesy Deutsches Museum, Munich.)

for example, was an alchemist only of the transcendental order, relating alchemy to religion but never having practised the art. T. Vaughan (Philalethes, fl. 1660) also took a mystical approach. He denied that bodies were formed by any mixture of elements, and he maintained with Helmont that fire was not an element. There was but one principle in the metals, namely mercury, and this was a derivative of water. Vaughan did try alchemical experiments, and his death is attributed to the inhalation of mercury vapor. Indeed, given the special role of mercury in alchemy generally, it might well have been the case that not a few adepts had suffered from the 'mad-hatter' syndrome.

The tendency of the seventeenth century 'scientific' philosophy to drive a wedge between man and the cosmos resulted in a revival of the alchemistic philosophy of cosmic unity. This esoteric cult of alchemists not concerned with making gold but with more basic questions of mystical and religious nature became known as Hermeticism, Hermes having been adopted as the figurehead. Alchemy became associated with various secret orders during this period, the Rosicrucians being a case in point, and the mystical aspect tended to become exaggerated. Robert Fludd (fl. 1620), one of the foremost English alchemical theorists of the first half of the seventeenth century, was both a supporter of Hermeticism and a Rosicrucian mystic, sharing this cult's concern for a new, chemically oriented learning in contrast to the ancient ways. For Fludd, light, darkness and water were more fundamental than the four elements of antiquity, since these 'secondary' elements are easily transmutable amongst themselves simply by condensation and rarefaction. The attribution of a special significance to water was not unique to Fludd, since a similar view was adopted by Van Helmont and others at that time.

THE DECLINE OF ALCHEMY

Newton (fl. 1690) was also influenced by Van Helmont as his extensive unpublished alchemical works show. There are indications that he was particularly interested in experiments with gold and mercury. His belief in transmutation seems to have become more restricted after about 1687, but there is little question that he continued to adopt an alchemical world view.

With the advance of the eighteenth century there was a decreased interest in alchemy, and in the wake of new chemical discoveries the older alchemy began to fade away. Yet in a sense alchemy never quite disappeared. There remained the eternal dream of the adept, and of the not-so-adept, to achieve real transmutation. Charlatans and those who treated alchemy as a means to gain riches had always been around, and this type of 'quack alchemy' survived the demise of the respectable art. In contrast to this residual physical alchemy, mystical alchemy became nearly totally associated with quasi-religious sects and cults. It was not until the nineteenth century clarification of atoms, elements and the periodic table, however, that the possibility of producing gold by chemical means could be said to have been scientifically disproved.

THE RISE OF CHEMISTRY

The immediate predecessors of Boyle and the chemists were the alchemists and iatrochemists, and without these preliminary advances in theory, experiment and techniques it is not possible to visualize the later developments. There are both differences and similarities. The chemist's goal was neither the transmutation of metals nor the preparation of medicine but a science of nature to be understood in its own right. Yet some aspects of chemical theory continue along the lines set down by the alchemists and iatrochemists. The ancient four-element theory underwent thorough revision. Boyle (fl. 1660) postulated as a definition of 'element' that which cannot be further decomposed. Although this was to a certain extent a case of *petitio principii*, it did help to clarify matters by allowing the metals to be classed as elements. And it contains the essence of the modern chemical definition of element. But while it provided a criterion by which to separate, distinguish and purify substances one from another, it left aside the more fundamental question of their structure. A deeper understanding of what this criterion really meant, distinguishing as it did homogeneous bodies from heterogeneous mixtures and compounds, had to await the introduction of atomic conceptions.

The iatrochemical sulfur—mercury—salt theory was considerably reformulated by Becher (fl. 1670), but it continued to apply to all substances and not just to the metals. The three principles became three earths, namely *terra pinguis* (fatty), *terra mercurialis* (mercurial), and *terra lapidea* (vitreous), where the first was essentially characterized by combustibility. Stahl (fl. 1690) elaborated this theory, introducing the term 'phlogiston' to designate the combustibility factor. The metals were considered compounds of metallic calx (oxide) and phlogiston. To obtain the metal it was then only necessary to heat the metallic calx with some substance rich in phlogiston such as coal or charcoal. Nearly another century passed before Lavoisier (fl. 1790) established the true nature of combustion and oxidation, which resulted in the demise of phlogiston and all forms of the sulfur—mercury theory of matter.

THE CHEMICAL ATOM

Meantime, however, it had been observed that substances combine with others in definite proportions. Not just any ratio could pertain between the components of water, for example, but it was found inevitably to consist of oxygen and hydrogen combined in the ratio of one part by weight of hydrogen to eight parts by weight of oxygen as H_2O. Not only was the proportion found to be constant but substances tended to consist of simple multiples of such a proportion. Hydrogen peroxide (H_2O_2), for example, has a ratio just twice that of water. The laws of constant proportion, multiple proportion and combining weights were fully accepted by the early nineteenth century.

A convenient explanation of these stoichiometric laws was provided at the turn of the century by the atomic theory of Dalton. According to this theory, all matter is composed of small atoms which cannot be further decomposed. Each kind of element was thought to have its own characteristic atoms, and hence the atoms of different elements have nothing in common with each other. Conversely, all the atoms of any one element were considered to be identical in their nature. And compounds were taken to consist of combinations of different kinds of atoms.

This refinement of Boyle's notion of an element, however, did not solve the question of structure but merely shifted it to another domain. Whether this fundamental question could be reconciled with the traditional problem of cosmic unity and *materia prima* remained to be seen. At first glance these might have appeared to stand in opposition to scientific principles.

PROUT'S HYPOTHESIS

Given this atomic theory of the chemical elements, it soon became essential to determine the relative weights of the different kinds of these atoms with respect to each other. One or the other having been adopted as a standard of reference, all the others could be measured quantitatively. Although hydrogen, taken as $H = 1$, served this function initially, it was later found more convenient to adopt the convention that the value for oxygen is a whole number ($O = 16$) whereby hydrogen then became $H = 1.008$.

That there must be some *materia prima* fundamental to all the elements had been adhered to even by Boyle, but the question of its nature had become more specific with the introduction of atomic conceptions. In this connection, Prout made two very pregnant observations about 1815. First, calculations of those atomic weights available at the time were nearly integral when hydrogen was taken as unity. Second, hydrogen is then very likely that of which all the elements are composed. If so, then hydrogen was the unifying principle of matter – the *materia prima* – and the atomic theory could be reconciled with the traditional belief in the cosmic unity.

While Prout's hypothesis initially met with a rather favorable reception, some atomic weight chemists, notably Berzelius, considered this opinion to be an affront to their laborious efforts to determine experimentally the atomic weights of many elements with an accuracy good to several decimal places. Alternatives were suggested to try to save this interesting hypothesis. In the 1840s Marignac proposed that one half an atom of hydrogen might fit the data better, and other sub-multiples of hydrogen were suggested about 1860 by Dumas. But such arbitrary fractional units of hydrogen were hardly suitable for ready acceptance as the *materia prima*, and the hypothesis became suspect.

4

Transmutation in the twentieth century

AT THE THRESHOLD

None of the goals of the older alchemy had really been achieved. The metals had not been perfected, and gold had not been produced. Although the medicinal preparations of the iatrochemists may have lengthened life, the alchemical elixir of immortality remained as elusive as the fountain of youth. No one had found the philosophers' stone, yet the dream of transmutation lingered on.

Historically we now stand at the threshold of success. In rapid succession transmutation was discovered in nature and brought under human control. How this was achieved will be dealt with sequentially in the next five sections, as the following overview indicates:

(1)	1896–1905	discovery of natural transmutation (1903)
(2)	1906–1919	transmutation artificially stimulated in laboratory (1917)
(3)	1920–1935	artificial radioactivity discovered (1934)
(4)	1936–1940	nuclear fission artificially stimulated in laboratory (1938)
(5)	1941–1981	transmutation processes brought under control

Several general points should be kept in mind. The transmutation of the atoms of one element into those of another does occur in nature and can be artificially stimulated. Far more important, however, is the release of energy which accompanies these processes of transmutation. Without question it can be said that the 'gold' of the newer alchemy is energy. These processes are mediated not by any substance or principle like the philosophers' stone but do depend upon very special methods of observation and production. And finally, the unity of matter becomes realized in

56

the enlarged sense that a small set of sub-atomic particles are found common to the atoms of all elements, and that even these *materia prima* particles, as well as the bonds holding them together, can be linked both by conversion and by conservation to this same 'gold'.

NATURAL TRANSMUTATION: 1896–1905

Introduction
Unbeknownst to scholars, alchemical adepts and scientists throughout all history, nature had been engaged in transmutation all along. The heavy elements uranium and thorium* were slowly converting themselves into somewhat lighter elements. No one even noticed nature's secret, because this was quite unexpected and the appropriate methods for observing this behavior were still lacking. The discovery of radioactivity and of electrons* at the end of the nineteenth century led to an understanding of natural transmutation just a few years later.

Electrons and matter
During the previous century it had been observed that the atomic weights of the elements are nearly unit multiples of hydrogen and can be ranked in regular periods according to their properties. In the tradition of *materia prima*, Crookes suggested that perhaps the elements are really quasi-

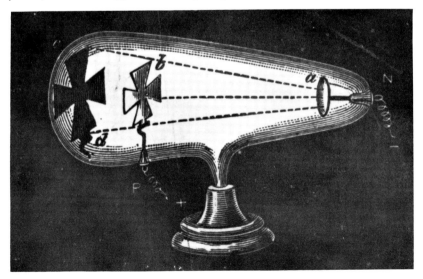

Figure 10. A Crookes tube.

compounds not in the chemical sense but of some primal 'protyle' consisting of very small particles. This suggestion was consistent with other
evidence that he had obtained while working on the phenomenon of
gaseous discharges. His 'radiant matter', otherwise known as cathode rays,
behaved like charged corpuscles, since they could be deflected from a
straight-line path by means of a magnet (see Fig. 10).

It was not until 1897, however, that J. J. Thomson definitely confirmed the corpuscular nature of cathode rays. These material corpuscles
were quickly linked with the unit of negative electricity already familiar
to electrochemists. And the discovery of these sub-atomic electrons had
implications for the idea of cosmic unity.

What was by no means clear at the time was whether these electrons
were material in the traditional sense. Perhaps their mass was only a property they possessed in virtue of their extremely high velocity, approaching that of light, and they were otherwise merely disembodied electrical
charges.

Radioactivity

Cathode rays proved to be one of the most fruitful scientific discoveries of
all time. In 1895 Röntgen discovered a secondary effect produced by
these rays as they strike the solid metal anode within a gas-discharge tube.
These X-rays were found to pass right through the glass wall just as light
would do. Also, they were found to induce fluorescence in certain sub-

Plate 6.   Antoine-Henri Becquerel. (From M. E. Weeks, *Discovery of the
Elements*, p. 776; courtesy American Chemical Society.)

stances and to darken a photographic plate even if metal obstacles were placed in the way. Becquerel began to examine particular substances to see if they also emitted X-rays. A few months later he discovered that these luminescent and photographic effects could be duplicated without any complicated electrical apparatus. Only a lump of uranium ore was required. But there was an important difference. The X-ray effect ceases once the discharge stops, whereas Becquerel's 'uranium radiation' continued unabated. His radioactivity* could not be turned off — it was a natural phenomenon new to history.

Radiations and radioactive substances
Becquerel had been able to duplicate effects found with discharge tubes, suggesting the existence of a very penetrating type of radiation analogous to the X-rays. This type of radioactive radiation was later identified by Villard, and it became known as gamma rays.* Giesel and others were soon able to isolate that part of Becquerel's radiation conforming to the cathode rays, and these became known as beta rays.* Yet a third type — alpha rays — were first noticed by Rutherford, because although they could barely penetrate a piece of tissue paper, they could strongly ionize the molecules of a gas as they passed through. These alpha rays were assumed at the time to be like soft X-rays, although they are now known to be helium nuclei.

Other substances were soon found which also emitted some of these same types of rays. Thorium compounds emitted not only radioactive radiations but also another 'emanation' which Rutherford found to resemble a gaseous substance. He was right, for 'emanation',* or radon, is the highest homologue of the chemically inactive or inert gases, helium, argon, neon, krypton and xenon, which had just been discovered in the 1890s, mostly by Ramsay.

The Curies in 1898 separated 'polonium' and 'radium' from the uranium ore, pitchblende, by chemical means. It was by no means clear at that time, of course, why these two elements should be co-present with uranium. If the existence of these new elements was of great interest, the most startling revelation was the intensity of their radiation. Radioactivity was thus as intense as it was persistent, but no one knew why.

Methodology
Radioactivity was discovered through luminescence and photography, and chemical methods were employed to separate the intensely active substances. Clues had been given by allied events occurring with electrical discharge phenomena, and the beta rays were found to be merely high-speed cathode rays somehow projected by radioactive substances. But it was with the alpha rays that radioactivity and gaseous ionization* formed

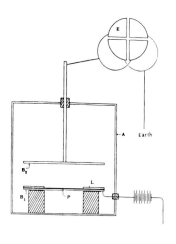

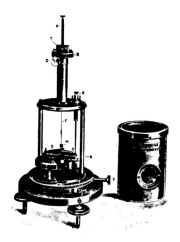

Figure 11. An electrical method of detection. (From T.J. Trenn, *The Self-Splitting Atom*, p. 29; courtesy Taylor & Francis Ltd.)

a particularly fruitful merger. Whatever their nature might be, these ionizing alpha rays could provide a convenient means to trace radioactive events. This was particularly relevant when chemical separations were concerned, for the quantities involved were miniscule. Alpha rays projected from radioactive substances strongly ionize the gas in their immediate vicinity. With suitable instrumentation such as an electroscope or

an electrometer the resultant ionic current can be quantitatively measured. This electrical method of detection is particularly sensitive to any
changes in radiation intensity such as might be induced by inserting a thin
metal foil in between or by changing the distance between the radioactive source and the detector. Other factors remaining constant, any
change in the ionic current would be attributed to a change in the intensity of the radiation emitted by the source and hence in its radioactivity.

Radioactive change
The discovery of radioactivity led to the discovery of natural transmutation. While it is now well known that radioactivity is associated with the
self-transmutation of one element into another by radioactive change, this
was not at all obvious at the time. Rutherford and Soddy made this pro-
found discovery during a period of joint research extending from late
1901 to early 1903. Extensive use was made of the electrical method to
trace the radiation intensity of substances separated by chemical methods.
They were able to separate from thorium compounds something which
they designated thorium-X and which exhibited distinct chemical properties. Thorium compounds seemed normally to have a constant radiation
intensity, if the curious radioactive emanation, thoron, could be kept
occluded within the mass of the compound. But with the chemical separation of thorium-X, the thorium compound appeared to have lost nearly
all its activity, whereas this new thorium-X seemed to possess nearly all
the activity originally associated with the thorium compound. In the first
instance this suggested nothing more than had the analogous case of
uranium-X which Crookes had separated from uranium compounds a
short time before. Furthermore, in both cases the substance designated
'X' was found to decrease in activity over the course of time, whereas the
original compound was found to have regained its activity. The rate of
change of activity was measured by the electrical method and the two
were found to be inversely proportional to each other. But there are
chemical processes such as that of π-bromonitrocamphor which also
exhibit this type of behavior, asymptotically approaching an equilibrium
condition, where the 'rise' and the 'decay' processes just balance each
other. And there are physical processes such as luminescence where the
intensity of radiation decreases exponentially as a function of time. It
thus remained to be proven that radioactivity was neither some special
type of 'luminescence' or allied effect induced from external sources, as
Becquerel, the Curies, Lord Kelvin and others believed, nor a chemical
process involving molecular synthesis or dissociation. Rutherford and
Soddy claimed that the process was nothing less than the *production* of
one substance by another, operative on the *atomic* level. Atoms of

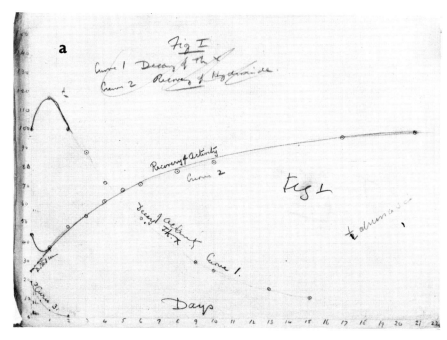

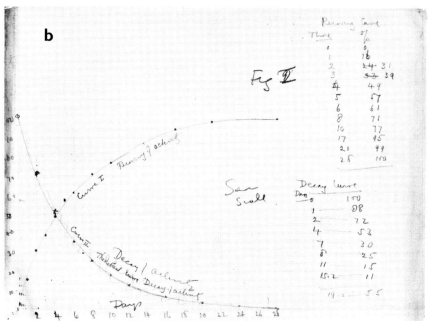

Figure 12. Decay and recovery curves from laboratory notebooks of Rutherford and Soddy (1902). (Courtesy Cambridge University Library.)

thorium were disintegrating and producing atoms of thorium-X, which in turn were producing atoms of the elusive emanation, and so forth.

At one stage of theory Rutherford and Soddy thought the daughter products were energized when they were produced, and that the decrease of activity over time was the fading away of this excited state (see Fig. 12). Everything fitted nicely into place, however, if the emission of radioactive rays was linked directly with the disintegration event, much like the other side of the same coin. But a theory is not necessarily true just because it is consistent and can account for all the known facts. If it were not for their firm experimental evidence that the processes occurred independently of factors such as the molecular state, extreme temperature variations, and all physical and chemical conditions generally, their case for natural trans-mutation would not have been nearly as strong as it was. Since a process was occurring which simulated the unimolecular reaction of chemistry yet which could be shown not to involve molecules or interatomic events, the correct conclusion was drawn that these were sub-atomic processes. And since these substances could be separated by chemical methods, albeit in unweighable quantities, traceable nevertheless by the electrical method, these must have distinct chemical properties. It is this spontane-ous production of atoms of one kind from atoms of another, at rates characteristic of the parent substance yet independent of any external influence or human control, that constitutes natural transmutation.

Transmutation directly observed
Radioactive change is thus a type of transmutation whereby the atoms of one element decay into somewhat lighter ones. The evidence in support of the theory of atomic disintegration was sound but intimately con-nected with experimental methods unfamiliar to chemists – and seeing is believing! After wrapping up the theoretical case for transmutation with Rutherford in Montreal, Soddy went to London to perform a spectro-scopic experiment with Ramsay, an expert in working with extremely small quantities of gases. It had meantime been established that uranium also gave off an 'emanation' similar to that of thorium and that the direct parent of this radon was Curie's new element radium. By good fortune small quantities of radium bromide had just become commercially avail-able through the efforts of Giesel working at the Buchler chemical plant in Brunswick. After liberating the occluded gaseous radon from a few milligrams of radium bromide by dissolving this salt in water, Soddy and Ramsay introduced this radon into an extremely fine spectrum tube which was then sealed off. After several days this tube, which had origin-ally contained only radon, began to yield the D_3 line characteristic of helium. It continued to grow brighter daily. This production of helium by the emanation of radium was the first case of natural transmutation to be established by methods previously known.

The energy of transmutation
Ramsay and Soddy also measured the actual volume occupied by the equilibrium quantity of radon from a known weight of radium under normal temperature and pressure, and thereby calculated what this would be for one gram of radium – hardly more than one cubic millimeter! But other experimenters had shown that one gram of radium evolves 100 calories per hour. Radium was known to take several thousand years to transmute completely into radon, and this intensity would gradually diminish in the course of time. But by the time this one gram of radium had completely decayed, it would have evolved over ten thousand calories, or a quantity of energy over four million times greater than that evolved by the same volume of hydrogen and oxygen when they explode to form water. This completely experimental result proved that the energy associated with radioactive change was of a magnitude hitherto unknown and quite outside the domain of chemistry. But what was the source of this enormous evolution of energy?

There were only two possibilities, but both seemed untenable. Either the atoms were being continuously replenished from some outside, perhaps cosmic, source or these atoms contained this vast energy stored in some fashion within their very structure. And if the latter alternative was perhaps surprising, it was at least not inconsistent with conservation principles.

Conclusion
Radioactive change as a type of transmutation occurring spontaneously in nature was observed and confirmed by means of a variety of appropriate methods. Two of the most characteristic radioactive phenomena are the persistence of the radiation and its intensity. Taken together with other evidence it became clear that quantities of energy were being evolved which were many orders of magnitude greater than anything previously known in nature. If we compare this natural transmutation or modern alchemy at this stage with that envisioned by the adepts of antiquity, several points of interest can be identified. The type of transmutation discovered was occurring spontaneously in nature independent of all physical and chemical conditions. Therefore no human effort could influence this process in any way, which contrasts with the expectations of the older alchemy. A change from one metal to another is involved, but its most characteristic feature is not a process of perfection but one of gradual deterioration into lead. The medium by which these observations were made was not the philosophers' stone but electrical, chemical and spectroscopic methods. And the most interesting product of such transmutation was not gold but energy, immaterial yet associated with the *materia prima*.

PERIODIC TABLE OF THE CHEMICAL ELEMENTS, 1920.

	Group O	Group I	Group II	Group III	Group IV	Group V	Group VI	Group VII	Group VIII
1 Hydrogen 1.008									
A	2 Helium He 3.99	3 Lithium Li 6.94	4 Beryllium Be 9.1	5 Boron B 11.0	6 Carbon C 12.00	7 Nitrogen N 14.01	8 Oxygen O 16.00	9 Fluorine F 19.0	
	10 Neon Ne 20.2	11 Sodium Na 23.00	12 Magnesium Mg 24.32	13 Aluminium Al 27.1	14 Silicon Si 28.3	15 Phosphorus P 31.04	16 Sulphur S 32.07	17 Chlorine Cl 35.46	
A	18 Argon A 39.88	19 Potassium K 39.10	20 Calcium Ca 40.07	21 Scandium Sc 44.1	22 Titanium Ti 48.1	23 Vanadium V 51.0	24 Chromium Cr 52.0	25 Manganese Mn 54.93	26 Iron Fe 55.84 · 27 Cobalt Co 58.97 · 28 Nickel Ni 58.68
B		29 Copper Cu 63.57	30 Zinc Zn 65.37	31 Gallium Ga 69.9	32 Germanium Ge 72.5	33 Arsenic As 74.96	34 Selenium Se 79.2	35 Bromine Br 79.92	
A	36 Krypton Kr 82.92	37 Rubidium Rb 85.45	38 Strontium Sr 87.63	39 Yttrium Yt 89.0	40 Zirconium Zr 90.6	41 Niobium Nb 93.5	42 Molybdenum Mo 96.0	43	44 Ruthenium Ru 101.7 · 45 Rhodium Rh 102.9 · 46 Palladium Pd 106.7
B		47 Silver Ag 107.88	48 Cadmium Cd 112.40	49 Indium In 114.8	50 Tin Sn 119.0	51 Antimony Sb 120.2	52 Tellurium Te 127.5	53 Iodine I 126.92	
A	54 Xenon Xe 130.2	55 Caesium Cs 132.81	56 Barium Ba 137.37	57 Lanthanum La 139.0	58 Cerium Ce 140.25	59 Praseodymium Pr 140.6	60 Neodymium Nd 144.3		61 · 62 Samarium Sa 150.4
		63 Europium Eu 152.0	64 Gadolinium Gd 157.3	65 Terbium Tb 159.2		66 Dysprosium Dy 162.5		67 Holmium 163.5	68 Erbium Er 167.7
		69 Thulium Tm 168.5	70 Ytterbium Yb 172.0	71 Lutecium Lu 174.0	72	73 Tantalum Ta 181.5	74 Tungsten W 184.0	75	
B		79 Gold Au 197.2	80 Mercury Hg 200.6	81 Thallium Tl 204.0	82 Lead Pb 207.20	83 Bismuth Bi 208.0	84 Polonium (210)	85	76 Osmium Os 190.9 · 77 Iridium Ir 193.1 · 78 Platinum Pt 195.2
A	86 Radium Emanation (222)	87	88 Radium Ra 226.0	89 Actinium	90 Thorium Th 232.12	91 Uranium X₂ (Brevium)	92 Uranium U 238.18		

Only the six spaces marked ——— are vacant places

The figures above the name of the Element are the Atomic Numbers, and those below the Atomic Weights.

Figure 13. Periodic table of 1920. (From T. J. Trenn, *Radioactivity and Atomic Theory*, p. 498; courtesy Taylor & Francis Ltd.)

ARTIFICIAL TRANSMUTATION: 1906–1919

Introduction
Once natural transmutation had been recognized, it was only a matter of time before the proper 'tools' and methodology became available to perform it under laboratory conditions. What was required was a method by which to influence or trigger a transmutation and appropriate means by which to observe the event. A proper trigger turned out to be alpha particles. Although alpha rays were still construed as soft X-rays in 1902, towards the end of that year they were found to be high speed particles several orders of magnitude more massive and energetic than electrons. Indirect evidence had linked these alpha particles with helium for several years, and this connection was confirmed in 1908. These massive alpha particles projected into the atoms of other substances might well be expected to trigger atomic disintegration. This was indeed expected, and the expectation was found to be warranted. Of course, alpha particles had been transmuting single atoms in this way for ages. The problem then became one of proving this under laboratory conditions.

Artificial transmutation attempted
Ramsay with Soddy had confirmed a case of natural transmutation whereby the highest known member of the inert gases, radon, was apparently degrading itself into the lowest member, helium. After this breakthrough, Ramsay began a series of researches designed to bring the natural process under human control. Using alpha radiation from radium emanation as his 'philosophers' stone', Ramsay obtained results that startled the scientific world of his day. He found that copper seemed to be degraded into the first member of its group, namely lithium, and that thorium could likewise be degraded to carbon. The radon itself did not always completely degenerate into helium but, in the presence of water, was decomposed only as far as neon, and in the presence of copper sulphate down further to argon. Since the *degree* of decomposition had apparently been moderated by the chemical conditions, Ramsay considered this to be evidence for controlled transmutation. None of these findings was substantiated by other researchers, and indeed all these experiments were found to be flawed especially by small gas leaks. But the failure of these particular experiments did not vitiate his basic view.

It was well known that water could be decomposed into hydrogen and oxygen in the presence of radium, and Ramsay with Cameron succeeded in synthesizing water from hydrogen and oxygen under the influence of another alpha emitter, radon. They also showed that carbon monoxide and hydrochloric acid vapor could be decomposed into their constituents. Ammonia was converted into nitrogen and hydrogen with their 'philo-

Plate 7. Sir William Ramsay. (From M. W. Travers, *A Life of Sir William Ramsay*, frontispiece; courtesy Edward Arnold (Publishers) Ltd.)

sophers' stone', then synthesized again from the component gases. These observations indicating molecular decomposition and synthesis were substantiated and extended by subsequent research. Ramsay and Cameron had observed that each particle of radon seemed to produce a discrete chemical effect, and they were correct in interpreting these molecular events as being caused specifically by the explosive emission of alpha particles. Although he did not identify the alpha particles with helium, Ramsay also speculated that these same alpha particles might be able to disrupt the atom too. While he had offered faulty experimental evidence in support of this view, his speculation that the alpha particle might be equally effective for atomic transmutation was eventually substantiated. It was not the idea as such but the particular choice of traditional chemical and spectroscopic methods that had doomed from the start his efforts to establish artificial atomic transmutation.

The problem of transmutation
During the first two decades of this century there was a marked increase in the general expectation that the process of atomic transmutation would soon be brought under human control. But it was not until 1919 that this age-old dream of man to manipulate the principles of matter was announced to the world. Radon with its energetic alpha projectiles was by no means the only possible agency under consideration. Some thought that X-rays might cause atoms to disintegrate, and possibly atomic dissociation could be induced by extremely high temperatures. The presence of neon and helium in old X-ray bulbs was misunderstood by some as indicating that even an electrical discharge could be useful as a 'philosophers' stone'.

The issue was not merely an academic one, for the discovery of natural transmutation had revealed in matter an enormous and previously unsuspected store of energy, compared with which all previously known sources of energy shrank into near insignificance. It was clearly seen by some that this energy supply must be brought 'on line' in the foreseeable future. But not all had the foresight exhibited in the following passage from 1912:

> Under existing conditions civilisation must come to an end the sooner, the more rapid its progress is and the greater the progress is . . ., for present day civilisation might be described as due to the supplementing of available supplies of natural energy by means of the accumulated stores of byegone ages. These stores may last out a century or two longer, but it is obvious that the more glorious the zenith attained the more swift and sure will be the decline. The only possible way of escape known to science is by the solution of the problem of transmutation . .
>
> Every year brings appreciably nearer the inevitable coming struggle for the possession and control of the primary sources of natural energy . . .
>
> Upon this issue . . . appears to hang the whole destiny of the race, spiritual, intellectual and aesthetic as well as physical or material. [Reference 6.]

Artificial transmutation attained
It is now known that atomic transmutation induced by alpha particle bombardment is a rather rare event, and that an observational method was required capable of locating the proverbial needle in the haystack. The methods employed by Ramsay would have required an astronomical number of individual transmutational events to permit a measurement. Not only was the number of events utterly insufficient to have been registered by his methods, but there were accompanying processes which completely swamped the rare transmutational event. Even the electrical method which had served so well for the discovery of natural transmutation would have been quite inappropriate. A method of detection was required capable of registering individual atomic events without the disturbing influence of other concomitant processes. The optical scintillation method proved to be uniquely suitable.

With a small research group in Manchester, Rutherford had begun a systematic examination of the structure of the atom in 1909 using alpha particles as probes and the scintillation screen as a standard technique. Rutherford had become aware by 1911 of the special case of secondary recoil atoms, where alpha particles impact upon light-weight atoms such as hydrogen and helium. The interaction between such minute nuclei would provide the maximum disturbance to the nuclear fields of force. A careful investigation of this special case was possible, since each product had a characteristic range and each particle could be registered by the scintillation method. Calculations by Darwin showed that such knock-on 'H' particles* in an atmosphere of hydrogen should be detected at distances over one meter, and Marsden experimentally observed these by 1913. But there seemed to be too many H particles reaching the scintillation screen, and in 1915 Marsden suggested that some of these might be expelled directly from the radioactive source just like alpha particles. After all, if radioactive substances emitted helium why not hydrogen too? Rutherford personally examined this exciting possibility. If true, this would have been the first evidence that particles other than electrons and helium nuclei were ejected by radioactive atoms. And if hydrogen were a nuclear constituent – the long sought 'positive electron' – then Prout's hypothesis would have been vindicated just a century later. Rutherford was able to show that Marsden's anomalous H particles were mostly due to hydrogen occluded in the source, and therefore still simply knock-ons. But he found an anomaly of his own which he recorded in his laboratory notebook on 9 November 1917.

The number of H particles tended to increase preferentially in the presence of nitrogen, some apparently coming directly from nitrogen. 'I am detecting and counting the lighter atoms set in motion by α-particles', Rutherford wrote to Bohr just one month later. 'I am also trying to break

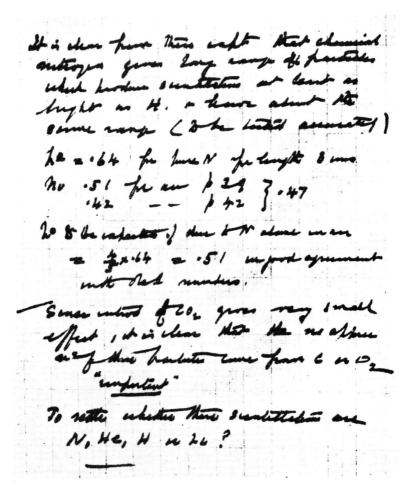

Figure 14. Data on H particles, a page from Rutherford's laboratory notebook recorded 9 November 1917. (Courtesy Cambridge University
Library.)

up the atom by this method. In one case, the results look promising but
a great deal of work will be required to make sure' (Reference 7). Presum-
ably Rutherford was referring to his nitrogen anomaly. While H particles
did not arise as a primary radiation from radioactive substances as
Marsden had surmised, they evidently could be kicked out as a secondary
radiation by impinging alpha particles – surely a far more interesting
possibility. 'I have got some rather startling results', he wrote to Bohr

about a year later, but he acknowledged the difficulties of 'getting *certain* proofs of my deductions' (Reference 8).

Part of the difficulty was that he had found apparent evidence for short range knock-ons from oxygen and nitrogen. It thus looked as if there were 'O' and 'N' particles in addition to the longer range 'H' particles. But oxygen and nitrogen atoms could not possibly have remained singly charged throughout the whole of their range in the gas as Rutherford had envisioned. Indeed, these seemingly special O and N particles turned out to be only very long range alpha particles. But if the O and N particles *could* have remained *singly* charged as they passed through the gas, then there was no reason why normal alpha particles could not have done the same. Rutherford admitted that he had been unable to distinguish the H particles arising from nitrogen disintegration from *possible* singly charged alpha particles, since their calculated ranges were nearly identical. If the H particles had been merely original alpha particles which had lost one of their two units of positive charge in passage, Rutherford would have had no evidence for artificial transmutation. He claimed that there was 'no obvious reason why nitrogen, of all the elements examined, should be the only one in which the passage of a swift α-particle led to the capture of a single electron' (Reference 9), and he was genuinely surprised a few years later to learn that alpha particles can become singly charged as they pass through a gas. On the basis of the evidence available, then, the anomalous particles observed in the case of nitrogen really could have been singly charged alpha particles. But they were not, and Rutherford had in fact succeeded in artificially disrupting nitrogen nuclei according to the reaction $^{14}_{7}N + ^{4}_{2}He \rightarrow ^{18}_{9}F \rightarrow ^{17}_{8}O + ^{1}_{1}H$. Yet it was 1919 before he published this first recorded case of artificial transmutation performed under laboratory conditions.

Materia prima revisited
Although this result was of fundamental significance, Rutherford was not particularly surprised that nitrogen should have been disintegrated under alpha bombardment. He had envisioned the nitrogen nucleus as consisting of a carbon nucleus with two hydrogen satellites which could be detached rather easily. Rutherford was astounded, however, that the alpha particle survived its impact with hydrogen. While the achievement of artificial transmutation was exciting, of even greater interest was the fact that hydrogen was definitely proven to be a fundamental constituent of nuclei. The tendency which seemed always to have existed 'to consider that all matter is composed of some one primordial substance' was particularly strong about 1915 (Reference 10).

Rutherford himself expressed greater interest in the fact that hydrogen, as the positive electron, was definitely proven to be a fundamental constituent of nuclei.

While it has long been known that helium is a product of the spontaneous transformation of some of the radioactive elements, the possibility of disintegrating the structure of stable atoms by artificial methods has been a matter of uncertainty. This is the first time that evidence has been obtained that hydrogen is one of the components . . . [Reference 11.]

Another commentator reported that the 'most important feature of Rutherford's fundamental discovery is that hydrogen is a primary constituent of the nucleus. This represents a return to Prout's hypothesis' (Reference 12).

Conclusion
One of the oldest of alchemical theories which postulated the existence of a fundamental substrate or prime matter has to some extent been vindicated by the discovery of structural units of atoms and nuclei, common to the elements of all substances. Differences among the elements could be reduced to the various possible combinations and relations of these basic principles. Transmutation, whether natural or artificial, is then basically the result of changes in any particular arrangement of these units.

Transmutation as control over natural processes had now been realized. These events were triggered by alpha particles playing the role of the 'philosophers' stone', and they were registered by a method of detection capable of recording individual events. This achievement has also brought us another step nearer to the *materia prima* units and to the promise of the 'new gold' – energy.

ARTIFICIAL RADIOACTIVITY: 1920–1935

Introduction
There are similarities and differences between natural transmutation as radioactive change and artificial transmutation. Radioactive change as the spontaneous and gradual degradation of elements such as uranium and thorium is a natural process, but so is the type of transmutation triggered by alpha particles. By suitable arrangements, whereby an alpha ray source is placed so that nitrogen or other atoms can be bombarded with these rays, this latter effect can be enhanced under laboratory conditions. It is in this sense that the transmutation of nitrogen differs from natural transmutation, for radioactive change cannot be influenced by varying the conditions. Both types of transmutation have in common that they require a special type of methodology for their detection. They differ in the type of particle ejected from the nucleus. Electrons and helium nuclei were

characteristic of radioactive change but hydrogen was expelled from nitrogen. All of this seemed to support the unity of matter and especially Prout's hypothesis. The next step in the scale of transmutation events is characterized by the stimulation of radioactivity and radioactive change in substances otherwise stable.

Isotopes: stable and unstable
Each element can be construed as a species, the individual members of which are atoms that can vary in their atomic weight within certain limits. The chemical properties are characteristic of the species but the atomic weight of a sample of any element is a function of the relative abundance of the particular varieties of that species. The existence of a range of varieties – the isotopes* – of any given element, and the fact that these are not all equally abundant, can lead to fluctuations in the value of the atomic weight measured from different samples of the same element. Since atomic weights are perforce only average values, this explains why these values are not exactly unit multiples of some basic units.

The existence of isotopes was proposed and verified by several workers, including Soddy and Moseley, about 1913. Of special concern here is that, although all of the varieties of any given element have identical chemical properties, they have physical differences over and beyond their weights. In particular, some varieties are stable or inactive while others are unstable and radioactive. As would be expected, the stable or near-stable varieties tend to predominate in nature, since these are the residue of a cosmic shaking out of inherent instabilities. All the atoms of the longest lived variety of uranium, $^{238}_{92}U$, are unstable, but it can be construed as a near-stable variety, since more than 10^9 years are required for one half of its atoms to become degraded to somewhat lighter atoms. The products of radioactive change do not in every case belong to the original species. When an atom of ^{238}U expels an alpha particle during radioactive change, this variety of uranium transmutes itself into an unstable variety of a quite different element, thorium. This reflects the loss of two units of positive charge and four units of mass associated with the expelled helium nuclei:

$$^{238}_{92}U \rightarrow {}^{4}_{2}He + {}^{234}_{90}Th$$

The helium nucleus has four units of mass, but only two of them are the fundamental positive unit found by Rutherford. The gain or loss of these charged protons must be compensated by a corresponding gain or loss of shell electrons, yielding a change of chemical properties; hence the shift from species 92 to 90. The change from 238 to 234 is only partially due to the loss of the two protons. The remaining two units lost reflects the existence of yet another nuclear constituent, massive like the proton but

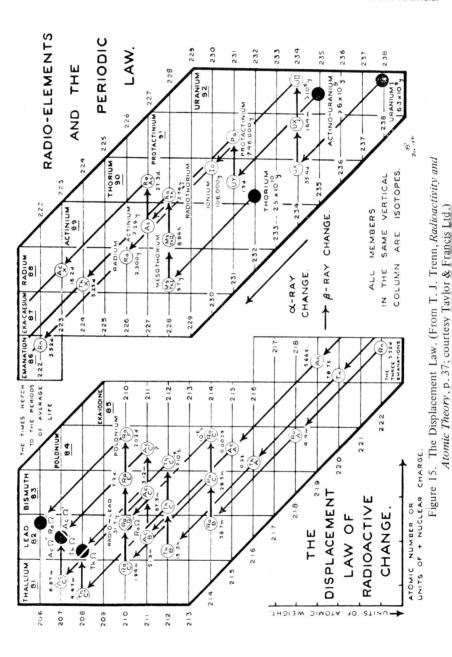

Figure 15. The Displacement Law. (From T. J. Trenn, *Radioactivity and Atomic Theory*, p. 37; courtesy Taylor & Francis Ltd.)

lacking any electrical charge. The varieties by weight that exist within any given species are the direct result of step-wise variations in the quantity of these 'neutrons' present in the nucleus of any given atom.

Neutrons*
Rutherford had already, in 1920, proposed the existence of a neutral constituent of nuclei, construed as a compound of negative electrons with his newly discovered protons. He envisaged this as a nucleus of unit mass and zero charge and hence able to move freely through matter and the shell electrons. Since it could more easily penetrate to the nuclear domain, such neutrons could be expected to combine with the nuclei of other elements or perhaps even disrupt them. It would thus be the ideal 'philosophers' stone'.

Now neutrons do exist and are an important constituent of all nuclei except hydrogen. They were first detected by Bothe and Becker in 1930 but not recognized for what they really were. They had begun systematic research on the emission of gamma radiation emitted from light-weight elements as a secondary effect of their bombardment by alpha particles. They detected a penetrating radiation from aluminum, boron and magnesium under alpha bombardment; but this was no cause for surprise, since these substances were known to yield protons as in the case of nitrogen. Yet it was curious why lithium and beryllium, which had been examined for proton emission without success, should emit this penetrating gamma radiation too. The rays from beryllium in particular were found to be much more intense than those exhibited by other elements, and they were of unparalleled penetrating power, being reduced to half intensity only after passing through nearly three centimeters of lead. Super-energetic photons were in vogue at the time, partly through the efforts of Millikan, who proselytized strongly in favor of the incorrect view that primary cosmic rays consisted of gamma rays of extremely high energy. Other researchers, including Curie and Joliot, repeated the results of Bothe and Becker and in 1931 found evidence for an elastic collision process between these very penetrating 'beryllium rays' and hydrogen nuclei, which then recoiled as fast radiant protons. They considered this to be further evidence that the beryllium rays were extremely energetic gamma ray photons of the Millikan type.

But in 1932 Chadwick examined the effect of beryllium radiation impinging upon the nuclei of various substances from hydrogen to argon. Given the laws of conservation of energy and momentum, he looked for a correlation in the energy transferred from the incident beryllium radiation to the target nuclei over the selected range of mass. While no consistent correlation emerged if beryllium radiation consisted of the then-popular photons, a definite correlation obtained on the

Plate 8. Sir James Chadwick. (Courtesy American Institute of Physics,
New York, Meggers Gallery of Nobel Laureates.)

assumption that the incident radiation consisted of neutral particles of
unit mass. Chadwick had thus discovered the neutron by direct experiment with various substances, and this suggested as a corollary that this
neutral particle was one of the constituents common to all nuclei generally.

Positrons

Chadwick's neutron differed from that of Rutherford in one important
respect. Rutherford's neutron was construed as a compound of the proton
and the negative electron, but Chadwick's neutron was slightly *more*
massive than the neutral atom of hydrogen. If the neutron is taken as the
fundamental 'protyle', perhaps the proton may be regarded as a 'compound' of a neutron with a positron – the true positive counterpart of the
negative electron. The existence of a particle with the same mass as the
electron but opposite in charge had been predicted in 1930 by Dirac.
Anderson in 1932 identified the 'electron pair' in photographic tracks
arising from experiments with cosmic rays, and Blackett independently
confirmed the existence of a pair of tracks curving away equally from a
common origin in opposite directions. Shortly thereafter, Anderson confirmed that such electron pairs could be produced by the passage of
gamma rays of sufficient energy through heavy elements.

Artificial radioactivity
Continuing their experiments with the Bothe–Becker effect, whereby alpha rays from polonium elicited in beryllium an unusually penetrating radiation, Irène Curie and Joliot in 1933–34 more than exonerated themselves for missing the neutron. The polonium source was placed behind a very thin metal window of aluminum through which the alpha rays had to pass before they could bombard the beryllium target. It was just part of their experimental set-up, but the aluminum was of course being bombarded with the alpha particles too. Joliot and Irène Curie noticed that the counter they were using continued to register impulses after the source had been removed. They traced this effect to radioactivity of the aluminum itself with a halflife of about 3 minutes. By dissolving the radioactive aluminum and then making a rapid chemical separation, they found that the radioactivity was actually associated with an isotope of the element phosphorus according to the nuclear reaction

$$^{27}_{13}\text{Al} + {}^{4}_{2}\text{He} \rightarrow {}^{30}_{15}\text{P} + {}^{1}_{0}n$$

where this isotope of phosphorus is unstable and decays to silicon with the emission of a positron according to the reaction

$$^{30}_{15}\text{P} \rightarrow {}^{30}_{14}\text{Si} + \beta^{+}$$

They also succeeded in transmuting boron and magnesium into artificially radioactive isotopes of nitrogen and silicon by this same radioactivation process.

It is of interest to note that aluminum foils had been used early on in the science of radioactivity. For example, Rutherford had distinguished alpha radiation from beta radiation by virtue of the great difference these exhibited in their power of penetrating aluminum foils. There can be little doubt that aluminum had been frequently transmuted into radioactive phosphorus during the succeeding decades. But this again reminds us of the importance of having the right 'philosophers' stone'. Fast positrons ionize just like electrons, but they quickly combine with free electrons and are annihilated within 10^{-10} seconds. Both the electrical method of detection and the scintillation method were more suitable for the detection of alpha radiation. Joliot and Curie used a Geiger-Müller counter which was extremely sensitive and could register individual ionizing events of the type involved. But this instrument was not invented until 1928, and it was several years before it became a standard laboratory instrument of the first rank. It is a case of a discovery awaiting the appropriate instrumentation.

Plate 9.   Enrico Fermi. (From M. E. Weeks, *Discovery of the Elements*,
p. 833; courtesy American Chemical Society.)

Fermi and neutron activation

Alpha particles were by no means the only 'philosophers' stone' available
which could transmute stable isotopes of one element into unstable isotopes of another. As had been expected since 1920, neutrons were
particularly effective in this regard. In 1934 Fermi, working with Segrè
and Amaldi, tried to produce effects similar to those reported by the
Joliot-Curies but using neutrons instead of alpha particles. A most convenient source of neutrons could be obtained from a sealed tube containing fine beryllium powder mixed with either about 100 milligrams of
radium sulphate or with an equivalent amount of radon. Fermi's group
thereby had a neutron flux of about 20 million neutrons per second with
an energy upper limit of about 10 MeV.

After trying several elements without success they succeeded in activat-
ing fluorine which then yielded a very short-lived beta activity according
to the nuclear reaction

$$^{19}_{9}F + ^{1}_{0}n \rightarrow ^{20}_{9}F \rightarrow ^{20}_{10}Ne + \beta^-$$

where the neon isotope is stable. The neutrons were being used as high
speed projectiles, and in this state they were more likely to disrupt the
nucleus than to be captured by it. Later that year Fermi and his co-
workers found that if the neutrons were first slowed down to thermal
velocities by paraffin or some other moderator, they were far more likely
to be captured and thus activate stable nuclei. Since it is not electrically

repelled, a slow neutron can spend a short time within the recipient nucleus, and there is thus a higher probability that it will be grafted into the system. Fermi's group was able to show that about 47 of the 68 elements tried were rendered radioactive by thermal neutrons,* evidently the most effective 'philosophers' stone' yet available.

Transmuting machines
Projectiles such as alpha particles and neutrons can perform transmutation by virtue of their high energy, whereby they interact and interfere with the nuclear structure, changing the balance between the nuclear constituents. If projectiles of even higher energy were available then greater feats of transmutation could be expected. It was with this goal in mind that various devices were designed which could yield projectiles of energy greater than provided by primary transmutation processes. These devices fall into two main categories depending upon whether they accelerate linearly or in a circle.

In 1932 Cockcroft and Walton prepared protons from a gaseous discharge tube containing an atmosphere of hydrogen, accelerated these by a voltage multiplier, and projected these high speed protons on to a lithium target. They obtained clear evidence of the production of helium according to the reaction

$$\,_1^1 H + \,_3^7 Li \rightarrow \,_2^4 He + \,_2^4 He$$

and this was probably the first case of artificial transmutation where the 'philosophers' stone' utilized had not originated from a primary transmutation event.

From 1931 another type of device, the cyclotron, was being developed by Lawrence and Livingstone, based on the principle of successively accelerating positively charged particles, but so altering the polarity that the particle spins in a spiral circle. Protons and deuterons (the nucleus of heavy hydrogen) were accelerated to such high energy levels that even before the end of the decade these projectiles made the ones previously available appear quite feeble. Even high energy neutrons could be produced by the cyclotron if accelerated deuterons were made to impinge upon a beryllium target, which then transmutes itself into a stable isotope of boron with the ejection of 9 MeV neutrons. Going the full circle, polonium (radium F = $\,_{84}^{210} Po$), which was the first radioactive substance to be discovered in 1898 by Marie Curie, was produced artificially in 1940 by bombarding ordinary bismuth, $\,_{83}^{209} Bi$, with deuterons of energy greater than 5 MeV produced by the cyclotron so that the proton is captured but the neutron escapes (Reference 13). That even the unstable isotopes which occur in nature could be produced artificially is perhaps but one overt sign of the ever increasing understanding and control that had been gained over nature's secret.

Conclusion

By this time there were several types of transmutation available and more than one way to perform the 'art'. Rutherford had succeeded in disrupting nitrogen nuclei, but the resulting isotope was stable. Joliot also used alpha particles, but the isotopes he obtained were unstable. Less than forty years after Becquerel had discovered an effect that could not be turned off, a way had thus been found whereby radioactivity could be turned on. Fermi found that neutrons, and especially thermal neutrons, were even more effective as activation agents. And with the arrival of accelerators, transmutation had begun to take on the character of a commonplace. But familiar as it had become, all successful transmutation to date was confined to the single atom scale. Individual nuclei were capturing incident projectiles, assimilating them, and then becoming transmuted into some variety of another species. And the particles ejected as a result were the familiar alpha particles, protons, and neutrons. Transmutation had come a long way but it still had far to go.

DISCOVERY OF FISSION: 1936–1940

Introduction

With the discovery that stable nuclei of one substance could be transmuted into unstable nuclei of other species, radioactive change was seen within a larger context of experience. Radioactivity could be induced in substances, but once started it could no longer be influenced by external factors. Once the difference was clear between stable and unstable varieties of an isotope, this observation became self-evident.

The neutron (1932) was evidently a basic constituent of atoms along with the proton (1917) and even the electron (1897). The modern *materia prima* had thereby reached one stage of completion. But there was no obvious reason why transmutation need be restricted to the expulsion of just one (neutrons, protons) or at most a few (alpha particles) of these units. Perhaps induced instability could lead to disruption of nuclei into much larger fragments.

Nuclear stability (see Fig. 16)

Although the nucleus consists mainly of protons and neutrons, not just any combination of these constituents is allowed. Except for the case of hydrogen, the nucleus of which is the proton itself, and some isotopic species (nuclides) in the lower part of the periodic table of the elements, the number of protons in any given nucleus never exceeds the number of neutrons. If the ratio of neutrons to protons is given as $n/p = R$ through-

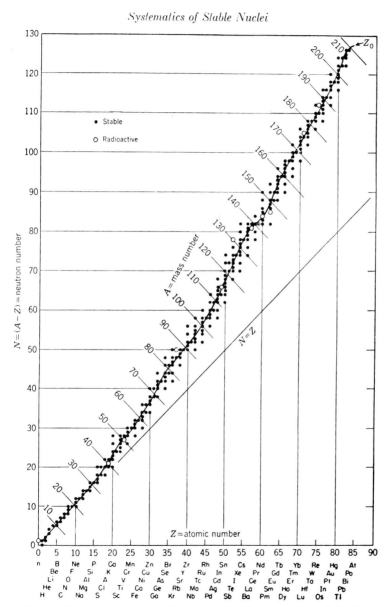

Figure 16. The path of stability: the naturally occurring nuclides for $Z \leqslant 83$. The solid line shows the course of Z_0, which is the bottom of the mass—energy valley, or the 'line of β stability'. (From R. D. Evans, *The Atomic Nucleus*, p. 287; courtesy the McGraw-Hill Book Co., Inc.)

out the periodic table, then allowable combinations are to be found only along a very thin path where the value of R deviates more and more from unity as a function of atomic weight, approaching $R = 1.6$ for the nuclei of the heaviest elements

All nuclei of elements heavier than bismuth are by nature unstable and are continuously undergoing natural transmutation. If an otherwise stable nucleus of some element, characterized by the number of protons — its p value — captures an alpha particle or a neutron, the amalgamation may well have an n/p ratio that no longer falls on this narrow path of stability. To get back on this path it must then undergo further adjustments. In the case of neutron capture, for example, the new nucleus produced can get back on track by undergoing beta decay.

Transuranic elements

That hydrogen is the lightest element can be explained by the fact that the proton is one of the basic constituents of all atoms as envisioned by Prout, Crookes and Rutherford. But why there should be a heaviest element and why this should be uranium was not at all obvious. Today we know that some transuranic elements may be artificially produced and that very likely these once existed during the formation of the universe. In the 1930s it seemed perfectly plausible that nuclei heavier than ^{238}U could be produced simply by adding enough neutrons.

Fermi's group continued their research with neutron activation by examining the case of uranium* and thorium.* Upon capturing a neutron, ^{238}U was expected to become ^{239}U, and likewise ^{232}Th to become ^{233}Th. If these then underwent beta decay, they would yield nuclei of substances shifted by one place in the periodic table. Specifically, according to the well-known displacement laws established in 1913 by Soddy, Fajans and others, these isotopic varieties of uranium and thorium would become nuclei of elements one place higher in the periodic table. Thorium ($^{233}_{90}$Th) would thus become a variety of protactinium ($^{233}_{91}$Pa), and uranium ($^{239}_{92}$U) would then have produced a variety of a new element 93. If so, then this would be the first case of artificial production of elements beyond uranium. Now there was nothing wrong with this theory. ^{239}U was certainly possible if ^{238}U could capture a neutron, and the stepwise beta decay of this 239 body should be expected to yield the transuranic elements 93, 94 and so forth. Fermi's group in Rome detected beta-decay isotopes having halflives of 10 seconds, 40 seconds and one of 13 minutes. Chemical separation methods seemed to suggest that the latter, construed as element 93, was a higher homologue of rhenium. Subsequent experiments by the Fermi group seemed to rule out lead, bismuth, and the elements 86 through 91, hence providing more indirect evidence that the expected transuranic elements were involved. But other researchers were unable to confirm the presence of such elements, and Fermi's claim for the discovery of transuranic elements (which had meantime been hailed as a victory for Italian Fascism) was discredited. In 1940 McMillan discovered that neutron activation of thin uranium foils yields a pair of beta emitters with halflives 23 minutes and 2.3 days, respectively. The first proved to be ^{239}U (discovered by Hahn) and the second an isotope of the new element 93, although chemical analysis done with Abelson showed that its properties were not similar to those of rhenium but to the rare earths. This neptunium ($^{239}_{93}$Np) does undergo further beta decay to form nuclei of element 94, plutonium* ($^{239}_{94}$Pu), as was confirmed in 1941 by Seaborg for the case $^{238}_{93}$Np to $^{238}_{94}$Pu. Fermi's group had had the right idea but their experimental results did not support it. The halflives were confused, and the isotopes they found seemed to have conflicting chemical properties. Something else was going on.

*Nuclear fission**
So intense was the interest in extending the periodic table beyond uranium that little attention was being paid at the time to whether or not other modes of transmutation might be involved with neutron activation. The clearest statement of caution was that of Ida Noddack (1896–1978), who in 1934 questioned the chemical results obtained by Fermi's group and felt it was unjustified to check for products only as far down the periodic table as lead or bismuth. She claimed that there was no reason to exclude types of nuclear reaction other than the familiar ones expelling protons and alpha particles. 'It would be plausible, that heavy nuclei would break up into several *larger* pieces under neutron bombardment, yielding isotopes of known elements that are not closely associated with that being irradiated' (Reference 14). While this may have been plausible, there was no clear evidence for such nuclear fission at the time. Thus during the period 1934–1938 it went unrecognized that nuclear fission had actually taken place in the samples used by Fermi's group yielding a variety of fission products in the middle range of elements from gallium to samarium. Several years, therefore, intervened before experimental evidence made the hypothesis of Noddack an inevitable conclusion.

An experimental clue to what was really taking place was provided by Curie-Joliot and Savitch in 1938. They activated uranium by means of neutron irradiation, and one of the products that they were able to separate seemed to resemble lanthanum chemically. It was initially thought to be a variety of actinium ($^{231}_{89}$Ac), supposed to arise from the alleged transuranic element ($^{239}_{93}$?) by two successive alpha-ray changes. The association of actinium with lanthanum was entirely reasonable, since it had long been known that actinium follows the reactions of lanthanum, element 57, very closely. To verify the actinium hypothesis Curie-Joliot and Savitch carefully precipitated the activity with lanthanum and

Plate 10. Ida Noddack, *née* Tacke. (Courtesy Deutsches Museum,
Munich.)

actinium carriers. Separation of the carriers indicated, unexpectedly, that
the activity remained with the lanthanum and not with the actinium! Two
conclusions follow immediately, although the full implications were
understood only in retrospect. If the activity did not go with the actinium
carrier then this was evidence against the production of its alleged grand-
parent, element 93. Conversely, if the activity went with the lanthanum
carrier then the new active product might be not just *like* lanthanum but
possibly lanthanum as such. If so then fission had occurred.

The experimental evidence which clinched the case for fission came
from the work of Hahn, Meitner and Strassmann in 1938. They were
looking for elements with atomic number higher than 92 from neutron
bombardment of uranium, and unbeknownst to themselves at the time
they achieved uranium fission. They activated a mass of uranium by
means of neutron irradiation and obtained three types of substance
which behaved as if they were isotopes of radium, actinium and thorium,
respectively. Assuming that ^{239}U had been produced, they might have
expected these to undergo two successive alpha-ray changes to varieties of
radium, as exemplified by $^{231}_{88}Ra$. In this particular case, further
transmutation by successive stages of beta decay would first yield $^{231}_{89}Ac$
followed by $^{231}_{90}Th$. But they had obtained *four* such alleged 'radium—

actinium—thorium' sequences simultaneously, and it is symptomatic of the theoretical difficulties that, faced with this experimental evidence, they did not even attempt to apply the displacement laws to these cases rigorously. It was towards the end of 1938 when they began to realize that they were really dealing with varieties of the elements barium, lanthanum and cerium and not their higher homologues. The decisive evidence was that their 'radium' could not be separated from the barium carrier but it could be separated from the radium carrier. Even so, Hahn phrased their conclusions with caution, since he considered nuclear fission to be still beyond ready acceptance at the time. Assuming that uranium nuclei had split into two large fragments, the one set being isotopes of cerium (58), lanthanum (57) and barium (56), they searched for the other 'halves'. Once on the trail they soon were able to identify the presence of strontium (38) and rubidium (37), along with the expected counterparts of fissioned uranium (92), viz. selenium (34), bromine (35) and krypton (36) among the fission products. This was incontrovertible evidence that uranium undergoes an entirely new type of transmutation in which the heaviest known elements divide or split into two approximately equal parts. Now the fission products are not necessarily paired exactly in the manner of the fission fragments. Whereas the atomic number of the fission fragments should total 92 in the case of uranium, e.g. selenium + cerium, bromine + lanthanum, or krypton + barium, these fragments may either give off their excess neutrons to become lighter varieties of the same element, or emit a sequence of beta rays to become different elements with nearly the same weight. In practice, both these sequences occur in such a way that there is still an excess of neutrons.

Secondary neutrons and energy
As soon as Hahn and Strassmann had published their evidence in 1939, the crucial importance of this discovery was very quickly recognized internationally. Meitner and Frisch had received an advance copy of the publication and concluded that such fission would be theoretically possible. It became readily apparent generally that such fission would release secondary neutrons and an abundance of energy.

Fermi for one immediately pointed out that the quantity of neutrons in the uranium nucleus was too great for all of them to be shared between the fission fragments. Since the n/p ratio is nearly 1.6 for uranium but about 1.3 for the middle range of the periodic table, several excess neutrons should theoretically be released from the nucleus undergoing fission before the two new nuclei — the fission products — could be brought within the path of stability.

Experiments soon indicated that large quantities of energy were released in fission, as evidenced by the fact that the heavy fragments move at very high speeds. Already by 1913 it was recognized that Einstein's mass–energy equivalence, $E = mc^2$, pertains to nuclear structure. Since energy must be applied to a system in order to separate its component parts and bring them to a dispersed equilibrium state, then the total *mass* of such an infinitely dispersed system is greater than the mass of that same system before it undergoes dispersal. Thus the mass of a nucleus is less than the total of each constituent's mass as measured when it is outside the nucleus. In other words, the conservation of mass-energy requires that this mass differential manifest itself as energy packed within the nucleus. The energy equivalent of one atomic mass unit can be calculated and turns out to be 931 MeV. Even radioactive change can yield alpha particles with an energy of nearly 9 MeV and hence cause a loss of nearly one hundredth of a unit of mass. By 1930 Aston, using mass spectrometry, had determined the mass defects of nearly all known isotopes. In the case of uranium fission, and that of similarly heavy elements, the mass defect between the original nucleus and the fission products totals about 22% of one amu and hence slightly over 200 MeV.

Conclusion
Not only had a new kind of transmutation been achieved but the large difference between the initial and final states increased the significance of the mass–energy equivalence. In the case of natural transmutation, radium disintegrated into two inert gases, namely radon and helium, liberating about 5 MeV of energy. In contrast, the energy yield in the case of nuclear fission is about forty times greater. The 'new gold' stakes had been increased, but the transmutational process still remained of the single atom type.

CONTROLLED TRANSMUTATION: 1941–1981

Introduction
If the first four decades were characterized by the discovery of transmutation and of a multitude of 'philosophers' stones', the four decades following the announcement of nuclear fission was a period during which the 'art' became better understood, exploited, and controlled. Perhaps the single most important factor that in retrospect can be seen as the watershed was the release of a flux of secondary neutrons as a sort of 'waste' product from the fission of heavy nuclei. If neutrons could be captured by heavy nuclei, causing them to become unstable, then in principle these excess neutrons should be able to do the same to other heavy nuclei in their immediate vicinity. This is the key which was to make the 'new gold' available, no longer just in fits and starts on the single atom scale, but in a sustained manner to be delivered either all at once or over a convenient period of time.

Sustained transmutation

The idea of a secondary effect caused by the debris from radioactive change was not really new. Back in the early days Rutherford and Soddy, for example, had looked for evidence that perhaps the alpha particles given out by thorium compounds might induce a secondary effect within these compounds. It was considered especially surprising to many that the halflife of radium remained unaffected by self-bombardment from its own energetic alpha particles. This general idea was vindicated with the discovery of artificial transmutation, but the 'nitrogen anomaly' was on a scale entirely different from that required to affect the rate of radio-active change of radium or of any other substance.

Neutrons, however, were a different sort of projectile, and in the case of uranium activation these were being released in small quantities in the direct vicinity of other heavy nuclei. Already in 1939 it had been experimentally confirmed by Joliot's group in France, among others, that the quantity of neutrons within a solution of activated uranium was greater than that provided by the irradiating neutron source. This confirmed not only the existence of this 'waste' neutron flux expected on theoretical grounds — about 2 or 3 per fission event — but also that this flux was sufficiently large that, even after neutron loss due to absorption, capture,* escape, and other causes, there remained a large enough aggregate excess sufficient perhaps to permit a self-sustaining reaction.

Now the mere existence of 'excess' neutrons by no means guarantees that a self-sustaining state of transmutation will be achieved. There are several competing factors which must be taken into consideration and optimized. When nuclei of heavy elements such as uranium are irradiated with neutrons, one of three events is likely to occur. The neutron may simply collide with the nucleus and be slowed down, it may be captured by the nucleus and become one unit heavier, or it may cause that nucleus to fission.

Most of the nuclei of uranium found in nature are of the variety $^{238}_{92}$U, its nearest competitor ($^{235}_{92}$U) having an abundance of only 0.72% compared with that of 99.28% for the former. Only fast neutrons with energies above 1 MeV can cause nuclei of ^{238}U to fission. If a sample of ^{238}U were irradiated by neutrons and fission were to occur, many of the excess neutrons would simply escape from the surface of the sample. Because of the relatively high inelastic scattering cross-section for nuclei of ^{238}U, many others would become slowed to thermal velocities. But slower neutrons tend to be captured by ^{238}U to form ^{239}U, which preferentially undergoes beta decay. Neutrons with intermediary energies tend to be captured, especially at certain specific 'resonance' energy levels. Hence, although most of the neutrons liberated by fission are energetic enough to cause ^{238}U to undergo further fission as a secondary process,

unless special arrangements are made, relatively few really do so. Therefore, the effective yield of 'waste' neutrons with ^{238}U alone is generally
insufficient to permit a fast chain reaction.*

The case of ^{235}U is entirely different, for it prefers thermal* neutrons,
but it can also be fissioned by higher energy neutrons. Nuclei of ^{235}U
have a relatively high cross-section for fission by thermal neutrons, and
this is the dominant mode of interaction. ^{235}U can produce enough
secondary neutrons to yield a slow chain reaction, provided that the
neutrons are reduced to thermal velocities, that a sufficient amount of
^{235}U is present in the sample, and that this neutron supply is not 'milked'
through neutron capture by the ^{238}U also present in the sample.

Neutron multiplication constant, or criticality factor
The crucial parameter in achieving a self-sustaining chain reaction is the
so-called neutron multiplication factor k. The gains and losses for each
generation, as with populations in biology, can total less than unity, more
than unity, or exactly balance. To maintain a population at stagnation
level there must be at least one birth on the average for each death,
throughout each successive generation. If the population is to grow, there
must on the average be more than one birth for each death. Similarly,
unless the average number of neutrons available for fission remains at least
slightly in excess of unity, throughout subsequent generations, a self-
sustaining reaction will stagnate and cease.

Nuclear reactors*
It is possible to control the value of k so that it does not exceed unity by
more than a very small fraction. With reactors it is neither necessary nor
desirable to have a high percentage of the fissile* isotope whether ^{235}U
or ^{239}Pu, but only enough to insure a sufficient excess of secondary
neutrons. To insure a sufficient quantity of neutrons of the right (low)
energy it is necessary to slow the neutrons down before they again come
into contact with the uranium. Excessive resonance capture of neutrons
by ^{238}U is an ever-present problem in the intermediate range of neutron
energies. The solution to this dilemma was ingenious. If the neutrons
escaping from a small mass of uranium are made to pass through graphite
or even water, they become slowed to thermal velocities, and can then be
permitted to enter another similar small mass of uranium. Given a three-
dimensional lattice array whereby nuggets of uranium are uniformly
separated from one another in a graphite medium, the secondary neutrons
would be 'moderated'* as they pass from one nugget to another. It was
with just this sort of 'pile' that Fermi's group in Chicago achieved the first
self-sustaining nuclear chain reaction in December 1942. Control of the
neutron flux was achieved by cadmium rods partially inserted into the

lattice. If fully inserted these would reduce k well below the critical value. But there should be some specific depth at which k will be just ever so slightly more than unity, and it is at this level that the nuclear fission process is both self-sustained and controlled. A small percentage of delayed neutrons aid in reactor control, for these delayed ones tend to pace the chain reaction with an overall average neutron life of about 0.06 s compared to a mean life of about 10^{-3} s for the prompt neutrons alone. To avoid possible confusion with the case of fission paced by slower neutrons, it might be mentioned parenthetically that when these 'delayed' neutrons are given off, they are initially as 'fast' as the prompt ones.

As the original supply of ^{235}U gradually becomes fissioned into products from the middle range of the periodic table, the control rods will gradually have to be withdrawn to maintain this critical value of k. But about once a year it becomes necessary to replenish the supply of fissile material.

Nuclear reactors which depend upon fission from neutrons that have been moderated to thermal velocities are designated thermal reactors. It is also possible to maintain a controlled chain reaction utilizing fast neutrons provided there is a proper (low) quantity of fissile material, in this case on the order of 10% of the total. Plutonium can be used as the fissile material in place of ^{235}U. Some of the neutrons with energies exceeding 1 MeV can cause ^{238}U nuclei to undergo fission. Others are resonance-captured to form more plutonium following the double beta decay of $^{239}_{92}$U to $^{239}_{93}$Np and then to $^{239}_{94}$Pu. Enough thermal neutrons resulting from collision with the lighter nuclei will remain to insure that the fissile material, whether ^{235}U or ^{239}Pu, will continue to undergo fission and keep k at a self-sustaining value. Nuclear reactors which also make use of fast neutrons are designated fast reactors.

The atomic bomb (see Fig. 17)
If the neutron multiplication factor k is allowed to increase well beyond unity then under certain circumstances it is theoretically possible that a fast chain will set in and accordingly a large number of the nuclei in a given sample will undergo fission in a very short period of time. In the case of uranium and other heavy nuclei, each fission event yields about 200 MeV of energy, or about 3.2×10^{-4} ergs. But the total number of uranium atoms in 1 g is Avogadro's number divided by the atomic mass of each atom, hence $6.024 \times 10^{23}/235$ or about 2.5×10^{21}, so that just 1 kg of uranium, if completely fissioned, would theoretically release about 5.0×10^{32} eV. Now according to $E = mc^2$, the energy equivalent of 1 g is 8.99×10^{20} ergs or 5.6×10^{32} eV. Thus the complete fission of 1 kg of uranium would theoretically release a quantity of energy of about 8.0×10^{20} ergs, that is, nearly equivalent to the complete energy-conversion of a mass of 1 g!

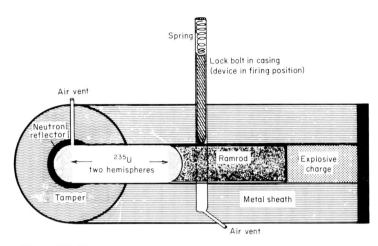

Figure 17. Figurative sketch of a nuclear explosive device. The high flux of prompt neutrons from a super-critical ^{235}U augmented by neutron reflection could induce and propagate a fast chain reaction, enhanced by photofission,* throughout the entire mass of purified uranium continuing even after this had begun to melt and expand. Once a fast chain reaction gets started, it is propagated about 100,000 times more rapidly than one paced by thermal neutrons, as long as there is a sufficient supply of fast neutrons. The parts of the bomb would not disperse in time to stop the reaction by self-destruction before a significant portion of the purified uranium would have undergone fission. (Note that a nuclear reactor – even a 'fast reactor' – must operate on the 'slow track', utilizing a chain reaction *paced* by slower neutrons so that even if it went out of control the result would at worst be a minor explosion but not the 'fast track' process characteristic of an atomic bomb.)

The key to this type of transmutation is that a sufficient number of fast secondary neutrons must be available to quickly fission other nuclei so as to initiate a 'fast track' chain reaction.

^{238}U is more a hinderance than an asset for the rapid exponential increase in k, although extensive fission of ^{238}U is possible, especially, for example, if it is saturated suddenly with a very high external flux of fast neutrons of energy greater than 1 MeV. To remove this from the ^{235}U presents a serious obstacle. Not only is ^{238}U by far the main variety of uranium but also the mass difference between these is barely more than 1%. Nevertheless this technical problem was overcome at great expense and effort by gaseous diffusion and by electromagnetic analysis, both methods exploiting the slight mass differential to the fullest. These techniques permitted ^{235}U to be separated from ^{238}U to the extent that samples containing over 90% ^{235}U were available by the kilogram. Such highly enriched samples are appropriate for some kinds of reactors but principally for bombs.

Removal of the unwanted ^{238}U from the fissile ^{235}U, while a necessary condition, is not sufficient to yield an uncontrolled fast chain reaction of the type known as an 'atomic bomb'. The configuration of the ^{235}U sample is also important, since enough neutrons must remain in the sample. Given a spherically shaped sample of ^{235}U it is intuitively evident that there is some critical radius of that sphere at which just enough energetic neutrons will remain within the sample to bring k up to unity. This value is known as the critical* mass. A sample of ^{235}U with a radius greater than this critical value would allow an uncontrolled chain reaction to set in. But as k increases, the high temperatures produced would soon gasify the material causing it to disperse and hence immediately bring k below the critical value. In order to perform this type of transmutation in a successful manner, whereby at least a significant portion of the ^{235}U actually undergoes fission before the sample turns itself off by self-dispersal, it is necessary to find some means by which a mass of ^{235}U with radius significantly larger than the critical value can be formed instantaneously and maintained in that state for a short time. Unless it is formed extremely rapidly, a premature chain reaction will set in causing melting or dispersion before enough fission 'generations' have taken place.

One method is that two or more sub-critical samples are hurled together by an explosive mechanism of some kind in such a way that these form just such a super-critical mass, but the more general practice involves implosion compressing the mass into a critical state. The more rapidly this can be done, the greater the chance will be that no premature chain reaction will begin, tending to oppose the formation of the super-critical mass. Ideally this super-critical mass should remain in this state until all ^{235}U nuclei have fissioned, which may take a significant fraction of a second. But by its very nature this is a self-destructing system, and hence the ideal cannot in principle be attained. Inefficient as this transmutational process would be from the point of view of the balance between the degree of fission of the super-critical sample and its tendency to self-destruct, if even just 1% of the ^{235}U in 1 kg were to fission successfully then the energy released instantaneously in this case would still be about 8×10^{18} ergs, or about 2.0×10^5 times the energy yield from an equivalent weight of dynamite. (Trinitrotoluene (TNT) yields 910 to 1085 gram calories per gram of TNT depending upon the packing density, and 980 may be adopted as a convenient average value. Since 1 gram calorie (mean) is 4.186 joules (absolute), the energy released from 1 kg of TNT can well be estimated as 4.1×10^6 J, or 4.1×10^{13} ergs.)

Plutonium ($^{239}_{94}$Pu) can also be used in place of ^{235}U, and the first successful test with a device of this type, exploded on 16 July 1945, allegedly contained several kilgrams of plutonium.* It was reported to have released an energy equivalent to about 20,000 tons of TNT, which upon conversion is about 18 million kg of TNT for that device. Since 1 kg of TNT releases about 4.1×10^{13} ergs, this nuclear device must have had an energy yield of 7.4×10^{20} ergs or 7.4×10^{13} J. But the maximum theoretical yield from 1 kg of uranium or plutonium, assuming it to be completely fissioned, is about 8.0×10^{20} ergs or 8.0×10^{13} J. Thus, assuming that this preliminary device functioned with, say, only 5% efficiency, about 20 kg of plutonium would have been required.

During subsequent decades many similar devices were tested with significantly greater yields, partly the result of greater efficiency and partly the result of larger quantities of uranium and plutonium. The use of larger sub-critical masses can improve the efficiency of the fission process as can better techniques for bringing about the critical state, but there are mechanical and other limits to these latter factors. Even much greater energy yields, up to one thousand times that of the original device, were achieved by using an atomic bomb as a trigger to detonate a thermo-nuclear bomb which transmutes by fusion due to D–D or D–T reactions, yielding a dense flux of fast neutrons and gamma rays to fission violently the ^{238}U casing, releasing energy in addition to the fusion energy.

Reactors cannot be bombs
Even if it went out of control, a nuclear reactor cannot simulate an atomic bomb. In the case of a thermal reactor, the quantity of fissile material, whether ^{235}U or ^{239}Pu, is only a small percentage of the total, and the physical arrangement is not conducive to maintaining the value of k above unity once this fuel began to melt, gasify and disperse. In the case of a fast reactor, there is a mass of ^{238}U in the core and, in the case of a fast breeder, also a blanket of ^{238}U surrounding the core. But the percentage of fissile material in the core, while greater than with thermal reactors, is gener-ally only of the order of 10%, insufficient to cause the onset of a fast chain reaction, particularly since it is not even in the compact form required of a critical mass* for it to go 'fast track'. The surrounding blanket, of course, gradually accumulates a fissile component of $^{239}_{94}$Pu which must be periodically removed to prevent a critical state from developing in the blanket. In general, however, the lack of concentration of the fissile material in the core, and with proper precautions also in the blanket, precludes such criticality from occurring. Even if an unwanted critical state were to be reached, the process would fizzle due to self-dispersal before even a slow chain reaction could really develop.

The techniques used to produce a *nuclear explosion* (i.e., an essentially instantaneous, self-perpetuating nuclear chain reaction) are very complex. A nuclear explosion must utilize a high energy neutron spectrum (fast neutrons). This results basically from the fact that, to produce an effective explosion the chain reaction must increase as rapidly as possible, utilizing the high energy neutrons produced in the fission reaction. The process by which a neutron is degraded in energy is time-consuming and largely eliminates the possibility of an explosion. (This also explains why power reactors that operate with a slow or thermal neutron spectrum cannot undergo a nuclear explosion, even if the worst accident is imagined. In the case of reactors that operate with higher energy neutrons, a nuclear explosion is also precluded because of the geometrical arrangement of the fissionable material and the rearrangement of this material if an accident occurs.)

The explosive ingredients of fission weapons are limited to plutonium-239, uranium-235, and uranium-233, because these are the only nuclides that are reasonably long-lived, capable of being produced in significant quantities, and also capable of undergoing fission with neutrons of all energies – from essentially zero or thermal to high energies. Other nuclides – as, for example, uranium-238 or thorium-232 – can undergo fission with high energy neutrons, but not with those of lower energy. It is not possible to produce a self-sustaining chain reaction with these nuclides, since an insufficient fraction of the neutrons produced in the fission reaction has sufficient energy to induce, and hence perpetuate, the fission reaction in these nuclides. [Reference 15.]

Nuclear reactors might be said to perform the most sophisticated type of transmutation yet achieved, with neutrons playing the role of the 'philosophers' stone'. It was accordingly with some considerable surprise that the scientific world learned in 1972 that there is at least one case of a nuclear reactor which had been operating about 1740 million years ago as a quite natural phenomenon in the geological past. Uranium samples found in a deposit at Oklo in Gabon exhibited an unusually low average concentration of ^{235}U. This and other findings are explicable in terms of continual fission of this ^{235}U over the course of time until there was no longer a sufficient quantity to maintain k at a value slightly in excess of unity. The natural transmutation discovered just seventy years earlier was mere child's play compared with this newly found type of 'natural' transmutation.

Transmutation machines

Although secondary neutrons from nuclear fission are a powerful philosophers' stone, there are other types as well. The charged particles and ions can be accelerated to very high energies, and the state-of-the-art of the various machines such as linear accelerators and cyclotrons have undergone considerable improvement over the years. The development of the high-energy machines used for particle physics began in 1946 with the discovery of various types of synchrotrons based upon the principle of phase-stable acceleration. Besides such variable-field devices, the microtron is a fixed-magnetic-field relativistic particle accelerator. Invented by Veksler in the 1940s, it can be designed with sector-focusing to produce electron currents in the 10 to 40 MeV range.

With the synchrotron, if the acceleration is brought about by fixed frequencies, the practical maximum energy of the particles is about 25 MeV. But the synchrocyclotron is capable of much higher energies. With the addition of alternating gradient focusing after 1952, energies in excess of 1 BeV were made possible. The machine at CERN can accelerate protons to 30 BeV, but this has recently been overtaken by other even more powerful atom smashers.

Transmutation controlled

The control of transmutation has both technical and social aspects. On the technical side, the 'art' of transmutation was developed with ever increasing variety and refinement in the form of nuclear devices (atomic bombs), nuclear reactors, and other transmutational machinery. But if all this and more is possible the need to bring these developments under control gradually became clear.

One of the first successful efforts was that regarding the rather flagrant testing of nuclear devices following initial military application in 1945. For more than a decade thereafter a great number of such devices of various sorts and of ever larger yields were detonated releasing vast quantities of radioactive debris into the environment. In 1957 an international group of scientists met in Pugwash, Nova Scotia, to 'assess the perils' which the further development of these weapons presented for humanity. This was but one of several such meetings of groups at that time greatly concerned about bringing the weapons side of nuclear transmutation under human control. Towards the end of 1958 the period of indiscriminate testing had come to an end. Similar efforts continue today, for new types of weapons utilizing these transmutational principles are continually being developed.

Nuclear power* for peaceful purposes, especially in the form of nuclear reactors to produce energy, has only recently come to be a subject of public concern. Attention has been drawn to the difficulties of safely storing nuclear 'waste', the possibility, albeit unlikely, of a reactor going completely out of control (even if it cannot simulate an atomic bomb), and the breeding of plutonium, which because it is so long lived is either a curse or a blessing for future generations, depending upon whether it is viewed as a waste product to be contained or as a wonderful new source of nuclear fuel adequate to meet future needs.

Human and social control mechanisms have begun to play an important role in deciding the future of transmutation. Who is to practise the 'art', how it is to be performed, and with what purpose, are issues which seem to have emerged as the new criteria.

Conclusion

Energy is the 'new gold' of modern transmutation, and it can be released in tremendous quantities either instantaneously or over long periods of time, for either destructive or for peaceful purposes. After several decades of discovery whereby one type of transmutation and philosophers' stone after another became available, the period since 1939 was marked by exploitation of these new-found possibilities. The achievement of ever greater technical control over the 'art' has become matched over the course of time by the implementation of social control mechanisms, introducing factors of the first magnitude in determining the future of transmutation.

5

Future expectations

INTRODUCTION

Now that the problem of transmutation has been solved during the first half of this century and that various modes of this 'art' were developed over the next quarter century, the center of interest has shifted to questions concerning the future. Discovery and invention still have their place but the dominant issues relate to the control and utilization of transmutation. Having become adept, we now possess a virtually limitless supply of the new gold – energy – at our disposal. Shall it be used for death and destruction, for peace and prosperity, for pure research, or not at all? In the following sections just a few of the relevant questions and issues will be touched upon, providing a general overview. Entire books and articles have been written on any one or more of these matters, and it is hoped that this introduction will be just that – an introduction. Let there, however, be no mistake about it. With the solution to the problem of transmutation our generation and those of the future are faced with a challenge unique to history. There is no turning back or passing the buck. The issues must be faced, and now! What was true in 1912 is today only that much more pressing, for 'the whole destiny of the race' hangs in the balance.

THE NEW GOLD

Energy is a commodity like any other but it is also the engine of productivity. With the solution to the problem of transmutation, therefore, the potential wealth of nations has become increased manyfold. It could not have arrived at a more appropriate time in world history. Civilization is dependent upon an adequate energy supply of some sort and has recently become very dependent upon coal and oil which are in limited supply and cannot be replaced. It took some considerable time to learn the simple fact that hydrocarbon reserves are finite, but it is now self-evident that it is only a matter of time before this unique inheritance runs low. But with recognition and understanding comes responsibility. Future generations will rightly condemn ours if we squander what remains of this precious material by simply burning it as a crude fuel. The less developed countries which as yet have no access to the new gold may be excused. Not so, however, those countries with credible alternatives. What little there remains of the 'black gold' should be doled out carefully to those in need of a quick energy 'fix', and should be used in the future as feedstock for the chemical industry, for transportation as required, as a starting point for protein food for animals and man, and for many other more refined purposes. In the interests of conservation, of the less fortunate of our generation, and of future generations as a whole, it thus becomes imperative to make as much use as possible of the new gold.

TRANSMUTATION NOW

Although it is perhaps not yet readily apparent to the general public, there is a pressing need for putting transmutation 'on line', and the sooner the better. There is a pattern in the past and we stand at a crucial transition point. The older fuels — firewood, animal wastes, coal, oil and gas — have each been used in succession with one becoming predominant for a time before being succeeded by another. The older ones do not become completely exhausted but become of lesser significance. This pattern has been repeated in the past with regular turnover intervals of about a century or less. The same could well happen with the fission reactor which is currently available in several varieties. Probably it will have to compete and perhaps even be phased out in the long term in favor of transmutation by fusion.* But the only viable alternative for our generation and probably for the next is the fission reactor. The energy gap cannot be eliminated by conservation measures alone, since conservation to our generation of enlightened adepts must mean very drastically reducing our own consumption of the world's supply of hydrocarbon fuels in favor of the needy of today and of the refinements of tomorrow. To continue the present policy and just 'cut back' consumption solves nothing but only postpones the 'inevitable conflict' for a little while longer. Transmutation is now available to many developed countries. These may elect not to accept this option for a variety of reasons. But failure to excercise this option dare not be construed as a warrant to continue to plunder at will the stores which should be set aside for the future and for those underdeveloped countries for which it is not yet a viable option. Not to bring transmutation 'on line' now through wilful procrastination and indecision must be the very height of irresponsibility.

TRANSMUTATION AND 'CATCH 22'

Transmutation is available and this fact must influence the outcome of the 'inevitable struggle' for the last remaining hydrocarbon reserves. Those countries which have meantime become thoroughly 'adept', even though some of these may traditionally be classed as 'less developed countries', will thereby have that much greater capacity for adaptability and be among the 'fittest' to survive. Those developed countries which had this option but failed to excercise it whether by design or by default will be that much less adaptable. And to the extent that some of these also find it necessary or convenient to continue to plunder the world's last reserves in a vain effort to stay afloat, they will incur the just wrath of the more enlightened nations. Having voluntarily placed themselves in a subordinate position, those developed countries which found it too difficult or inappropriate to become fully adept on their own may well be provided with the necessary facilities by those thoroughly adept and hence superior nations. Therefore, because of the checks and balances within the global system, *not* to embrace transmutation and the new gold turns out not even to be a viable option at all. To avoid 'going nuclear' is to become voluntarily underdeveloped. To continue to plunder global reserves to boot cannot be justified. To avoid embracing transmutation now is thus both unjust and nationally debilitating, and this only postpones the day when nuclear reactors must be accepted even by those who tried to avoid doing so. Transmutation is therefore really the only 'option', for it will be put 'on line' globally in the long run, where appropriate, whether voluntarily or otherwise.

Correctives will be required, especially for those underdeveloped countries which continue to use decentralized energy systems either because they have never had any other viable option or because they have adopted such a system as part of a degenerate political and economic process. The high costs of such decentralized energy systems will lead to a further weakening of these formerly developed nations and make it all the more imperative that they ultimately be drawn into the global systems of transmutation.

VOLUNTARY TRANSMUTATION

Since the coming of applied transmutation is inevitable, the best approach is to phase it in both nationally and globally with forethought and planning. It is imperative to plan ahead, since there is necessarily a long lead time between design and implementation. If nuclear power were not permitted to develop progressively but had to be stepped up within a few years in a crash program, there would inevitably be some cutting of corners, just to meet a world energy famine which by then would have become sufficiently imminent to be recognized even by the most stubborn of critics. Nuclear fission is still predominantly a source of electricity, but in time it could supply heat as well. Approximately one third of the total world's fuel consumption today is used by conventional power stations to produce electricity. If nuclear reactors were brought 'on line' in adequate numbers in tandem with alternatives such as hydropower the global consumption of conventional fuels could be reduced by one third. And there is no good reason why this capacity cannot be further increased or why electricity cannot be utilized for heating and other purposes such as mass transportation.

But it is not only the number of reactors which must be taken into consideration in bringing transmutation 'on line' but the type of fission reactor. There are two basic types, namely thermal and fast reactors, depending upon whether the neutrons are moderated to thermal velocities before they cause further fission events or whether the fast neutrons are permitted to participate in the fission sequence as well. Thermal reactors can fission ^{235}U but not ^{238}U. Fast reactors can fission ^{238}U and ^{235}U, and some of the fast neutrons are inevitably slowed by collisions to thermal velocities. But these two types can be distinguished in another way. Thermal reactors primarily consume fuel. Fast reactors can be designed to produce more fuel than they consume, and are hence known as 'breeder'* reactors. These are frequently designated 'fast breeders'* to imply both aspects, namely that they can use fast neutrons to slowly breed more fuel than they consume. Thermal reactors also breed some fissile material in the core, but this is not the function for which they were designed and the product is still considered 'waste'. In principle core breeders and blanket breeders can both yield a fresh supply of fissile material, although the reprocessing* required to extract the fissile material must be different.

FISSILE AND FERTILE* FUEL

There are large deposits of uranium (U_3O_8) throughout the globe, but nevertheless the supply is limited. Less than 1% of uranium is of the fissile variety, ^{235}U. There are other fissile fuels, namely ^{233}U and ^{239}Pu, but these do not exist in nature, and must first be bred. The price of uranium has already climbed to over $40 per pound, and even if mining techniques could keep up with the demand for a time, there are limits both in costs and in availability. Hence, if nuclear power is to be more than a transient phenomenon, it is necessary to increase the supply of fissile fuel. There is also the matter of efficiency. Thermal reactors utilize only about 1% of the fuel. On the other hand, fast reactors can already produce about 50 times more energy per fuel unit, and in principle they can be designed to utilize nearly 100% of the fuel. Thus, if only thermal reactors are used without reprocessing, the supply of fissile fuel would run short probably by the end of this century. One solution is to intentionally breed fissile ^{239}Pu from fertile ^{238}U and (also in thermal breeders) fissile ^{233}U from fertile ^{232}Th. Not only can fast breeder reactors make much more efficient use of the fuel available in the core but also — and this may be their decisive advantage over thermal reactors — there is a higher net production of neutrons in the fissile material, which are then available to convert fertile material to fresh fissile material.

Thus, in the interests of conservation and efficiency, it is imperative to plan now to phase in fast breeder reactors, or near-breeder reactors of the CANDU type, as a supplement to thermal reactors. Reprocessing of the fuel may possibly become standard practice for breeder and thermal reactors alike, although this can be avoided if near-breeders are utilized.

BREEDER REACTORS

A nuclear reactor, in order to function, must maintain a self-sustained reaction and hence have enough fissile material to maintain k slightly above unity. But over the course of time, the fissile material will have undergone transmutation by fission and will have to be replaced. But in principle, it would be possible to design the reactor such that while this batch of fissile material was undergoing transmutation, the available neutron flux would be made use of to convert non-fissile varieties of uranium or thorium into fissile varieties. If the conversion ratio is greater than unity, that is, if the reactor can produce at least as much fissile material of one variety as it consumes of this or another variety, then the reactor is a breeder. It not only has a high energy yield from the consumption of the primary fuel, but it produces as a by-product a quantity of fuel at least as great as that it consumes. The primary fuel consumption occurs in the core of the reactor. The breeding generally occurs in a blanket of non-fissile (fertile) uranium or thorium surrounding the core. Thermal reactors are not too conducive to blanket breeding, since slow neutrons cannot efficiently enter the blanket. But a fast reactor, without moderator and hence also utilizing the fast neutrons, is quite capable of this secondary breeding action. The primary fuel may be either ^{235}U, ^{239}Pu or

^{233}U on the order of 10–20% mixed with ^{238}U. The blanket can be either ^{238}U or ^{232}Th. In the first case the product is ^{239}Pu, and a given quantity of ^{238}U will be completely transformed to ^{239}Pu in a matter of less than two weeks. In the case of thorium ($^{232}_{90}$Th), the product is fissile ^{233}U and the conversion takes place within 5 or 6 months. In either case, the fissile fuel bred in this manner can be used as primary fuel. If properly functioning and with a net positive output, the breeder reactors can effectively increase the amount of fissile fuel beyond that provided by nature and produce an adequate supply for future generations.

By their very nature such breeder reactors tend to accumulate fissile material in the blanket, and this must not be allowed to reach the critical state. Therefore care must be taken to remove the blanket periodically and to reprocess the fuel.

THE THORIUM CYCLE

Thorium is present in nature both in its own right and as the daughter of fossil ^{236}U. As a result the abundance of $^{232}_{90}$Th exceeds that of uranium (^{238}U) by nearly a factor of four. Thorium is not fissile by nature, but in a situation such as within a reactor operating with ^{235}U or ^{239}Pu, the thorium (^{232}Th) will be bombarded with neutrons and upon capture form ^{233}Th. But ^{233}Th beta decays to protactinium ($^{233}_{91}$Pa) with a halflife of about 22 minutes, which means full transformation within 2 hours. The latter transmutes itself by a second beta decay into uranium (^{233}U) at a rate such that the first half is converted within one month. Thus, within five months a given supply of ^{233}Th will have become transformed into ^{233}U which is fissile and has a halflife of over 10^5 years.

The great advantage here is that with a slightly different reactor design, it is possible to develop a self-sufficient thorium cycle. Once the thorium has begun to breed ^{233}U it is clear that this fissile uranium can repeat the process. It is no longer necessary to have an initial input of fissile ^{235}U or ^{239}Pu. In other words, the thorium breeder can be made a self-sufficient system. The thorium cycle can be used either with thermal 'core' breeding or with fast neutron blanket breeding. Reactor types which can make use of the thorium cycle are still in the research stage. But once it is made commercially feasible, this achievement would provide an adequate supply of fuel for the best part of the coming millennium.

PLUTONIUM, WASTE AND CONSERVATION

Plutonium generally refers to the fissile variety of that element, $^{239}_{94}$Pu, produced as a by-product of the fast breeder. Its counterpart in the thorium cycle is ^{233}U which has no special name but is equally fissile and, like plutonium, either a curse or a blessing depending upon whether it is considered a 'waste' product or a fuel dividend. Plutonium can be bred on purpose in a fast reactor but it is also a by-product from thermal reactors. Although ^{238}U cannot be fissioned by thermal neutrons it can capture thermal neutrons to form plutonium. Hence, plutonium is an inevitable product of putting transmutation 'on line' and is not uniquely associated with fast* breeder reactors. Some varieties of plutonium could be used in thermal reactors as the fissile component instead of ^{235}U. But all varieties can be used to fuel fast reactors, and this is its special virtue. Plutonium in the core is incinerated. The amount bred in the blanket could, in principle, be so controlled that the net gain was either greater than unity, less than unity (as in thermal reactors) or exactly unity. If exactly as much plutonium was produced as incinerated then there could be a constant supply of new fuel without any undue proliferation of this substance beyond that deemed necessary for weapons. In the event that new reactors were being put 'on line', the supply of plutonium bred could be adjusted beyond unity so that there would be a net gain and hence a supply sufficient to fuel also these new facilities. It would then be economically advisable to strike a balance between the amount of plutonium produced and that incinerated in such a way that the entire system of facilities is taken into consideration and that the net effect would match the energy requirements of society without relying excessively upon mining to obtain fissile material. Plutonium would accordingly be construed not as a waste product to be disposed of but as a fuel dividend helping to conserve available natural resources.

Waste products do arise in fission reactors, for the core tends to accumulate fission products in the mid-range of the periodic table – most notably strontium ($^{90}_{38}$Sr) and cesium ($^{137}_{55}$Cs). Some waste products have relatively short lifetimes and can be disposed of by time alone. Others can be vitrified in unleachable glass and stored or otherwise disposed of in some appropriate manner. Recycling of even these 'waste' products may ultimately win the day. Meantime, adequate means are being found to take care of this waste. Perhaps one day the complete solution will be found, but it would be foolish to stop reactor research and development until this supposedly ideal approach is achieved.

TRANSMUTATION 'ON LINE'

One of the biggest difficulties in planning to get transmutation on line is insuring sufficient lead time. There is a minimum number of years, say 8 to 10, from start to finish. Given the rapid pace of social and political developments in our day and age, this is a relatively long lead time indeed. There are considerable design problems, for example, involved in making the breeder reactor feasible. It is therefore important not to hold up research and development, but on the contrary to continue research at full throttle and even put pilot research reactors 'on line' so as to continually improve the design by experience. The worst possible situation would be a sudden demand from a public finally awakened to its real predicament and thus requiring many fission reactors, both thermal and breeder, without adequate time for further research and safety considerations.

Optimistic estimates are that the fast breeder will be introduced progressively only after 1990 and that the full benefits of this fast reactor technology, on which serious research had begun about 1950, will not be realized until about 2040! This is a lag of some 90 years. On this same scale, if transmutation by fusion, whether in the form of fusion reactors or in the form of solar energy, is now in the early stages of research, it would be optimistic to expect that fusion could yield significant benefits much before the latter half of the twenty-first century. This is not to say that fusion will not have its day. Very likely solar (fusion) energy and fusion reactors will replace fission reactors in the long run much as transmutation by fission must displace the current overdependence on and misuse of hydrocarbon reserves in the short run.

FUSION TRANSMUTATION

The various types of transmutation accomplished to date have in common that they typically involve the disruption of one type of nucleus to produce another. It is transmutation by disintegration. But quite a different type of transmutation exists in nature and is the means by which stars generate heat and light. This is the fusion of the lightest elements, especially the heavy varieties of hydrogen, into compound nuclei which subsequently explode with the release of energy. Solar energy is essentially the energy of fusion from an extraterrestrial source, and a great deal of research is being conducted to convert this into a useable supply for the future. Thermonuclear fusion has been achieved on earth in the form of a bomb, but such an uncontrolled reaction is not useable as a steady source of energy. Controlled thermonuclear fusion, if it could be achieved, would be the ideal way to generate energy in the long term. The raw material is water, that is the small percentage of heavy hydrogen (deuterium) – about one ^{2}H in every 7000 atoms of hydrogen – and tritium, ^{3}H, even rarer in nature but which can be produced from ^{6}Li + n. One liter of water has the energy equivalent of, say, 300 liters of high grade gasoline. The problem is how to tap this supply.

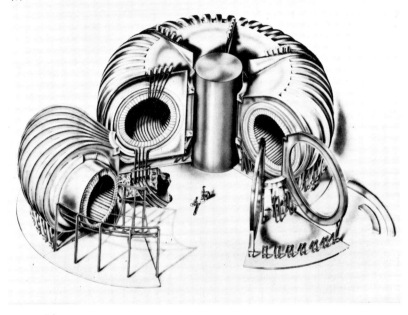

Figure 18. The modern Ouroboros: fusion toroids. (a) Artist's conception
of a fusion power reactor. (Courtesy Oak Ridge National Laboratory.)
(b) Overhead view of Scyllac, the major new experiment in controlled
thermonuclear research at the Los Alamos Scientific Laboratory.
(Courtesy American Institute of Physics, Niels Bohr Library.)

To achieve fusion it is necessary to heat a small quantity of deuterium gas to a temperature of about 100 million °C to get the nuclei moving fast enough so that they can fuse. But it is also necessary to contain the resulting hot plasma for durations perhaps as long as one second. Now, no material can support such temperatures (the Sun's surface temperature is less than 6000 °C) for that long without vaporizing. Ingenious attempts have been made to contain the plasma in a 'magnetic bottle' to insulate the hydrogen gas at temperatures required for fusion. But besides containment there is also the problem of producing these temperatures in the first instance. This can be approached in several ways, including the injection of energetic neutral atoms into the plasma which pass unimpeded through the magnetic field, are ionized and stopped, giving their energy to the plasma. A similar effect can be achieved by radio frequency excitation or by passing a current of several thousand amperes through the plasma. The major problem is to confine the hot plasma long enough for fusion to occur.

One of the most promising methods of plasma containment is that done using the Tokamak technique developed in the Soviet Union. Several years ago this technique yielded relatively quiescent plasmas of 10 or 20 million °C for durations of several milliseconds. For fusion reactors, however, it is estimated that the duration must be extended perhaps by as much as a factor of one thousand. More recently plasma temperatures in the region of 60 million °C have been attained.

Even more promising is the use of the Tokamak principle in conjunction with a modified version of the Stellarator, developed originally in America during the 1950s, as recent results in Germany have shown. This improved Stellarator can produce fusion energy without interruption, dispensing as it does with the pulsating *internal* magnetic field required by a Tokamak, the latter being utilized here merely to pre-heat the plasma confined by the helical external magnetic field alone.

In contrast to such closed systems, open-ended cylindrical devices exist which can 'pinch' the plasma with a strong magnetic field and confine it by magnetic 'mirrors' at either end. The long straight center section of such tandem-mirror fusion devices offers an ideal surface area for interfacing the cooling blanket with the power station's steam turbines.

Inertial confinement is another promising alternative method, releasing the energy available in fusion by irradiating small pellets of deuterium and tritium with intense laser light vaporizing the surface instantaneously. The expanding hot gas exerts such a pressure on the pellet that its density and hence temperature is increased to that required for thermonuclear fusion. The power requirement for the laser pulse is enormous.

In general, transmutation by fusion is still at the stage where more energy has to be put in than can be obtained from the system. It is hoped that before the end of this century fusion transmutation will have been achieved under laboratory conditions despite the net energy loss. If this were attained, it would be roughly analogous to the achievement of artificial transmutation by Rutherford in 1919 or perhaps that of nuclear fission by Strassmann and Hahn in 1938. Optimistic estimates indicate that a net energy gain may be available from fusion transmutation not before we are well into the twenty-first century.

HYBRID REACTORS

A fusion device is also a net source of fast neutrons and could hence be combined with fission reactor systems in at least two different ways. The best results would be expected with a straight cylinder fusion device at the core, say 10 meters long and 30 centimeters in diameter, as this would provide an adequate surface area. If the neutron flux emerging from this central cylinder was first thermalized and then captured within a sheath of natural uranium (mainly ^{238}U) surrounding this core, the resulting ^{239}U would yield plutonium (^{239}Pu) in due course, much like in the ^{238}U blanket of a fast breeder reactor. On the other hand, if the neutron flux were not initially thermalized before it entered the ^{238}U sheath, an amplification process would ensue which could provide, say, three times as many fast neutrons, as a secondary yield, on the outside of the sheath as were impinging at the core—sheath interface. If these neutrons were then moderated by means of a thermalizing layer of carbon, this augmented flux of slow neutrons could then be used to breed either ^{239}Pu or ^{233}U depending upon whether the 'skin' surrounding the carbon layer were uranium oxide or thorium oxide. Were this 'skin' prepared in a colloidal state, the resulting fissile material could be all the more easily separated and prepared for utilization in fission reactors. A suitably *thin* sheath of natural uranium only several centimeters thick would ensure sufficient neutron leakage to prevent a critical state from arising.

CONCLUSION

Only a few of the highlights have been touched upon in relation to what may be expected of transmutation in the future. Much has been made of late about new devices for destruction making use of transmutational principles. However, such uncontrolled uses should actually have no role in the enlightened future. As to the question of controlled transmutation, it will eventually come to be seen as the only viable option for the coming generation or two. At the moment it is difficult for many to see their way through the maze of issues. But there are several overriding facts which, if kept in mind, help to clarify matters. Civilization depends upon an adequate energy supply. Enlightened as it is and having solved the problem of transmutation it dare not refuse its benefits nor continue to plunder the storehouse of the poor and of the future. But if responsibility dictates that the world must wean itself from hydrocarbons now, and if fate dictates that transmutation by fusion is only possible in the long term, then there is *de facto* only one way to go, and that is to put fission transmutation 'on line' globally as soon as possible. But here the problem arises of achieving a balanced program taking into consideration the advantages and disadvantages of thermal and breeder reactors and the need to extend and expand the limited supply of fissile fuel for the future. Plutonium is but one of the many issues that arise and is often misunderstood, for far from being a waste product it is a new supply of fuel which conserves that provided by nature. If burned in a breeder reactor its production could be balanced in such a way that the net gain is just sufficient to keep a fleet of reactors running. But there is so much — so very much — misunderstanding concerning transmutation that one can only wish the veil of ignorance would quickly disappear. To us has been given 'the new pearl of great price' but unfortunately not the wisdom to recognize it for what it is. Perhaps there is yet time.

6

Reflections on transmutation

Transmutation turns out not to be a vain dream of the past but a vivid reality for the present and future. Its most characteristic features can be identified throughout the several stages of its history. Ancient alchemy may be seen in retrospect as the proto-idea, whereby the basic principles of matter would yield to human influence typically in the form of converting base metals into gold. Transmutation as change involving some *materia prima* was considered to be attainable through the medium of the philosophers' stone, and gold was to be the outcome. There is more than a little in common between this proto-idea and the reality of today. Modern transmutation is a change involving primary nuclear constituents which can be made to change their configurations. It takes one kind of philosophers' stone to observe such transmutation in nature and another kind, namely high speed particles and especially neutrons, to conduct the 'art' at will. The product of such change is the new gold – energy.

This is what Soddy had in mind when he pointed out in 1917, well in advance of its achievement, what the significance of this new kind of transmutation would be.

> If man ever achieves this further control over Nature, it is quite certain that the last thing he would want to do would be to turn lead or mercury into gold – *for the sake of gold*. The energy that would be liberated, if the control of these sub-atomic processes were as possible as is the control of ordinary chemical changes, such as combustion, would far exceed in importance and value the gold. Rather it would pay to transmute gold into silver or some base metal. [Reference 16.]

The history of alchemy up to the iatrochemical period can be seen as the main line of historical development leading to modern chemistry. But while transmutation and chemistry share a common concern for atomic mass, the nature and classification of the elements and other aspects of chemical thought, the point of divergence occurred during the first decade of this century. With the recognition by Rutherford and Soddy that the atom contained a store of energy within itself, having nothing to do with chemistry, it became only a matter of time before the source would be more clearly identified. It was 1911 when Rutherford proposed that the atom contained a central 'nucleus', and by 1913 it had become clear that this was the heart of the atom. Changes within the nucleus were producing parallel changes with the chemical electrons resulting in the shifting of elements back and forth across the periodic table according to the displacement laws of 1913. The nucleus was the source of those alpha, beta and gamma rays and of the neutrons, protons and positrons which were shortly to appear on the scene. And it was the nucleus which had stored within it the energy released both in radioactive change and in fission. But with this new understanding it became clear that transmutation belonged to a different domain than chemistry altogether. Chemistry concerns the interactions of atoms, molecules and larger aggregations, and to this extent it must deal with the electronic configuration beyond the nucleus. But interatomic energy is of an entirely different domain and magnitude than intra-atomic or nuclear energy. Transmutation concerns changes within the nucleus in the first instance, although these changes may be reflected by parallel changes in the chemical properties of the substance in question. Transmutation, therefore, is only indirectly related to chemistry proper, and it would be more appropriate to consider it trans-chemical.

Whatever its roots and origins, and however it may be classified, transmutation is the 'art' by which the new gold has become available to modern civilization. It is the hope and challenge of the future. Now that we have become adept, it is imperative that we learn how to apply the 'art' wisely for the mutual benefit of our own generation and of those to come. Only time will tell whether wisdom has prevailed.

Some historical data on transmutation

prehistoric		Oklo nuclear reactor
450 BC	Empedocles	Four-element theory of Greek philosophy
fl. 380 BC	Plato	*Timaeus*
fl. 340 BC	Aristotle	Prime matter
300 BC		Rise of metallurgical alchemy in Hellenistic Egypt
		Rise of medicinal alchemy in China
ca. 200 BC	Pseudo-Democritus	*Physica et Mystica*
fl. AD 175	Galen	Four-humor theory
fl. 300	Zosimus	Color sequence test
fl. 320	Ko Hung	Elixir of life
ca. 400		Zenith of Alexandrian alchemy
ca. 620		*Great Secrets of Alchemy*
ca. 750		Rise of Arabian alchemy
fl. 775	Geber	Sulfur–mercury theory
fl. 900	Rhazes	
fl. 1230	Albert the Great	
fl. 1250	Roger Bacon	
fl. 1280	Raymond Lully	
fl. 1280	Arnold of Villanova	
1300		Rise of Latin alchemy
fl. 1310	Pseudo-Geber	
ca. 1330	Petrus Bonus	*The New Pearl of Great Price*
15th c.		Alchemy embedded in esoteric symbolism
16th c.		Beginning of iatrochemistry
fl. 1530	Paracelsus	*Tria prima*
fl. 1530	Agricola	*De Re Metallica* 1556
fl. 1580	Libavius	*Alchymia* 1595 (textbook)
1600		Advance of iatrochemistry
1604	Basil Valentine	*Triumphal Chariot of Antimony*
1610	J. Boehme	Mystical alchemist
fl. 1620	R. Fludd	Alchemical theory

fl. 1620	J.B. van Helmont	Water the *materia prima*
17th c.		Transmutation considered a scientific fact
fl. 1640	J.R. Glauber	A true believer in alchemy
fl. 1650		Hermeticism
fl. 1660	R. Boyle	Element defined
		Sceptical Chemist 1661
fl. 1660	Philalethes	Mystical alchemy
fl. 1670	J. Becher	*Physicae subterraneae* 1667
		'Three earths' theory
		Sulphur–mercury–salt theory revised
fl. 1700	Stahl	Phlogiston
1775	Lavoisier	Oxygen
		Death blow to transmutation of metals
1808	Dalton	Atomic theory
1815	Prout	Prime matter is hydrogen 'protyle'
1819	Dulong and Petit	Specific heat related to atomic weight
1876	E. Goldstein	Kathodenstrahlen
1879	W. Crookes	Cathode rays as fourth state of matter
1895	Röntgen	X-rays discovered
1896	Becquerel	Natural radioactivity discovered
1897	J.J. Thomson	Electron discovered as corpuscle
1898	Curie	Polonium and radium discovered
1902	Rutherford	Alpha rays confirmed as particles
1902–3	Rutherford and	Natural radioactivity explained
	Soddy	Theory of atomic disintegration
1903	Soddy and Ramsay	Helium produced by radon
1908	Rutherford	Alpha particles confirmed as helium
1910–13	Soddy	Theory of isotopes
1911	Rutherford	'Nucleus' theory of atom
1911–13	Soddy	Displacement law
1912	Soddy	Energy crisis foreseen for the future
1912	Rutherford	Nucleus of atom confirmed
1913	Marsden	H particles observed
1913	Moseley	'Roll call' of the elements
1913		Geiger counter invented
1917–19	Rutherford	Artificial transmutation stimulated in laboratory
		Proton identified as Prout's 'protyle'
1919	Rutherford	Artificial transmutation reported
1919	Rutherford *et al.*	Start of period of transmutation by alpha particles
1920	Rutherford	Neutron predicted
1928		Geiger-Müller counter invented
1930	Dirac	Positron predicted
1930	Bothe and Becker	Neutron 'undiscovered'
1930	Aston	Mass defects determined
1931	Lawrence and Livingston	Cyclotron

1931	Curie and Joliot	Neutron 'undiscovered'
1932	Chadwick	Neutron discovered
1932		Start of proton-induced transmutation
1932	Anderson	Positron confirmed experimentally
1932	Cockcroft and Walton	Transmutation induced by artificially accelerated protons
1934	Fermi *et al.*	Neutron-induced transmutation begun
1934	Ida Noddack	Fission hypothesis announced
1934	Joliot and Curie	Artificial radioactivity discovered
1938	Curie and Savitch	'Lanthanum' connundrum
1938	Hahn and Strassmann	Nuclear fission discovered experimentally
1938/9	Meitner and Frisch	Fission theory
1939	Hahn and Strassmann	Published results on fission
1939	Joliot	Excess neutrons confirmed
1940		Start of control and application of transmutation processes
1940	Cork *et al.*	Artificial production of polonium
1940	McMillan and Abelson	Neptunium-93 discovered
1941	Seaborg *et al.*	Plutonium-94 confirmed
1942	Fermi *et al.*	First controlled chain reaction
1944	Seaborg *et al.*	Americium-95 discovered
1944	Seaborg *et al.*	Curium-96 discovered
1945		Atomic bomb detonated
1946		Synchrotron
ca. 1950		Start of research on 'fast' reactors
1950s		Fusion research with Stellarator
1952		Synchrocyclotron
1957		First Pugwash meeting
1958		End of indiscriminate nuclear testing
1960s		Fusion research with Tokamak begins
1972		Prehistoric Oklo reactor discovered
1973		The 'energy crisis' hits the headlines
1980		Positive results in fusion research with Stellarator and Tokamak

Glossary

Adept. Classical title of alchemists who had attained the secret of transmutation.

Alpha particle. The equivalent of helium nuclei expelled from the nucleus.

Beta ray. High speed electrons originating from radioactive changes.

Breeder. A reactor which produces excess fissile material while consuming other fissile material.

Capture. Reaction within the nucleus whereby it absorbs either an additional proton increasing its atomic number by one unit or an additional neutron remaining a member of the same species of element. In both cases its mass increases by one unit.

Chain reaction. A repetitive self-sustaining sequence of events each stage of which is triggered by the preceding, e.g. as in a critical mass where neutron fission releases secondary neutrons which cause further fission etc. In a reactor this chain reaction is paced by slower neutrons, even though fast neutrons can be involved, whereas in a bomb the chain is paced by faster neutrons.

Cloud chamber. A detection device yielding visible tracks from ionizing rays, invented by Wilson.

Critical mass. 'The mass of fissionable material required to produce a self-sustaining sequence of fission reactions in a system (a reactor, for example) is the *critical mass* for that system.

The chain of reactions will be self-sustaining if, on the average, the neutrons released in each fission event initiate [at least] one new fission event. The system is said to be *critical* when that condition exists . . . .

The escape probability becomes larger for smaller systems, inasmuch as the ratio of surface to volume increases as a system is made smaller. Thus, there is a *critical size* below which the chain reaction in a given system cannot be made self-sustaining.'†

In a bomb the fissile material is compacted in such a way as to permit the "*fast track*" chain reaction to occur, which depends upon the higher energy portion of the energy spectrum of prompt (10^{-14} seconds) neutrons and lasts only a few

---

† Charles D. Goodman, Critical Mass in *The Encyclopedia of Physics* (ed. R.M. Besançon), Reinhold, New York (1966), pp. 140–141.

microseconds for the entire sequence of fission generations. In a reactor the fissile material is distributed in such a way that no sufficiently compact state can possibly arise essential for this type of chain reaction. But the arrangement does permit a chain reaction that depends upon the lower energy portion of the spectrum and normally also upon delayed fast neutrons as well.

Electron. The quantum of negative electricity and the elementary particle bearing that charge.

Emanation. Generic name for the gaseous daughter products – radon, thoron, and actinon respectively.

Fast breeder. A reactor which converts fertile material into excess fissile fuel by the capture of neutrons and by operating at an average neutron energy of 0.1 MeV or greater.

Fast neutrons. In reactor theory, those neutrons with energy above the fission threshold in ^{238}U which is about 1.2 MeV. The average energy of fission neutrons whether prompt or delayed is about 2 MeV, although their initial energy can exceed 10 MeV.

Fertile fuel (fertile material). Non-fissile varieties of elements such as uranium and thorium which can be converted to fissile material by neutron capture, especially ^{238}U and ^{232}Th which can be converted to fissile ^{239}Pu and ^{233}U respectively.

Fissile material. Nuclear varieties which undergo fission by neutrons of all energies. In contrast to ^{235}U, ^{238}U can be fissioned only by fast neutrons.

Fission. The splitting of the nuclei of the heavier elements into two large fragments with the release of several excess neutrons.

Fission cross-section. The effective area measured in barns (10^{-24} cm^2) indicating the probability that particles or rays passing through fissionable material will induce fission.

Fissile nuclides have a very high cross-section for low-energy neutrons (^{235}U greater than 600 barns). ^{238}U has no cross-section in this energy range, but it shares with ^{235}U a similar low cross-section (0.5–1 barn versus 1–2 barn) for fast neutrons.

Fission products. The fission fragments after having adjusted themselves to the path of stability by expelling further neutrons, and by extensive beta decay.

Fusion. See *Nuclear fusion.*

Gamma rays. High energy photons.

Geiger-Müller counter. A detection device pioneered by Geiger and brought to fruition in 1928 by Walter Müller.

H particle. The designation for the proton and for the simple whole multiples of this unit up to ^{4}H = helium.

Ionization. The process by which an atom loses or gains one or more of its natural allotment of electrons and hence acquires a net electrical charge.

Isotope. A nuclear species the varieties of which all have the same atomic number but can vary in their neutron allotment within the range permitted by the path of stability.

Moderator. A substance such as water or graphite which can thermalize fast neutrons.

Neutron. An elementary particle with nearly unit mass and devoid of electrical charge.

Nuclear energy (nuclear power). The energy released by transmutation – the 'new gold'.

Nuclear Fusion. The union of two or more lighter nuclei to form a heavier nucleus with the liberation of energy.

Nucleus. The massive heart of the atom positively charged and, except for ordinary hydrogen, consisting of at least as many neutrons as protons.

Philosophers' stone. Supreme object of alchemy, as substance supposed to change other metals into gold.

Photofission. Fission of nuclei by energetic gamma rays exceeding a threshold of about 5.5 MeV for such isotopes as ^{238}U, ^{235}U, ^{232}Th, and ^{239}Pu. The fission fragments are strong sources of prompt gamma rays with energies up to about 7 MeV, although they average about 1 MeV.

Plutonium. Element 94 whose principal isotope $^{239}_{94}Pu$ is fissile and can be bred from $^{238}_{92}U$.

Proton. An elementary particle with unit mass and unit positive charge   .

Radioactivity. Spontaneous radioactive change with the emission of radioactive radiations.

Reactor. Usually refers to fission reactors as an assembly of nuclear fuel and associated apparatus for sustaining and controlling chain fissioning.

Reprocessing. The extraction of fissionable material from spent fuel by chemical and other methods to be reused in fission reactors.

Slow neutrons. Neutrons which have been slowed to intermediate energies (10^2 eV to 1 MeV) or thermalized to even lower energies by a moderator or by collision events without a moderator. ^{235}U is more easily fissioned by slow neutrons than by fast neutrons.

Sub-atomic change. Transformation involving changes within atoms.

Thermal neutrons. See *Slow neutrons.*

Thorium. Element 90, the most abundant variety of which is ^{232}Th; it can be made to yield ^{233}U by neutron capture.

Uranium. Element 92, the most abundant variety of which is ^{238}U; this can be made to yield ^{239}Pu by neutron capture. ^{235}U also occurs in nature normally in the approximate ratio of 0.7% 235 to 99.2% 238.

Appendix C

References

1. K.B. Hasselberg, 'The Nobel Chemistry Prize', *Les Prix Nobel en 1908*, Stockholm (1909), p. 21, following modern translation in *Nobel Lectures in Chemistry, 1901–1921*, Amsterdam (1966), pp. 126–127.
2. S. Brown, *Essays, Scientific and Literary* (1858) as cited in W. Ramsay, *Essays, Biographical and Chemical*, London (1908), pp. 21–22.
3. H. Schelenz, 'Der Hermetische Verschluss', *Chemiker Zeitung* (1907), pp. 339–340.
4. (a) H.S. Redgrove, *Alchemy: Ancient and Modern*, Rider & Son, London (1922) (2nd edn University Books Inc., NY, 1969), p. 82.
   (b) F.S. Taylor, *The Alchemists*, Schuman, New York (1949) (reprinted Arno Press, NY, 1976), pp. 175–176,
5. Reference 4(a), pp. 87–88; Reference 4(b), p. 187.
6. F. Soddy, 'Transmutation, the Vital Problem of the Future', *Scientia*, 11, 186–202 (especially pp. 200–202) (1912).
7. Letter from Rutherford to Bohr, 9 December 1917, in T.J. Trenn, 'The Justification of Transmutation', *Ambix*, 21, 53–77 (especially p. 71) (1974).
8. Letter from Rutherford to Bohr, 17 November 1918, in Reference 7, p. 72.
9. E. Rutherford, 'Nuclear Constitution of Atoms', *Proceedings of the Royal Society London*, 97A, 374–400 (especially p. 380) (1920); cf. Reference 7, p. 75.
10. W.D. Harkins and E.D. Wilson, 'Energy Relations Involved in the Formation of Complex Atoms', *Philosophical Magazine*, 30, 723 (1915).
11. Reference 9, p. 385; Reference 7, p. 74.
12. W. Lenz, 'Betrachtungen zu Rutherfords Versuchen über die Zerspaltbarkeit des Stickstoffkerns', *Naturwissenschaften*, 8, 181 (1920); cf. Reference 7, p. 74.
13. J.M. Cork, J. Halpern and H. Tatel, 'The Production of Radium E and Radium F (Polonium) from Bismuth', *Physical Review*, 57, 371–372 (1940).
14. 1. Noddack, 'Über das Element 93', *Angewandte Chemie*, 47, 653–655 (1934).
15. G.T. Seaborg, *Man-made Transuranium Elements*, Prentice-Hall, Englewood Cliffs (1963), pp. 60–61.
16. F. Soddy, 'The Evolution of Matter', *Aberdeen University Review* (1917), reprinted in *Science and Life: Aberdeen Addresses*, London (1920), p. 107.

Suggested reading

E. N. da C. Andrade, *The Atom and its Energy*, Bell, London (1947).
P. Beckmann, *The Health Hazards of NOT going Nuclear*, Golem, Boulder (1976).
M. P. Berthelot, *Les Origines de l'Alchimie*, Librairie des Sciences et des Arts, Paris (1885).
W. H. Brock, 'Studies in the History of Prout's Hypotheses', *Annals of Science*, **25**, 49–80, 127–137 (1969).
A. G. Debus, *The Chemical Philosophy: Paracelsian Science and Medicine in the Sixteenth and Seventeenth Centuries*, N. Watson, New York (1977) (2 vols.).
B. J. T. Dobbs, *The Foundations of Newton's Alchemy*, Cambridge University Press, Cambridge (1975).
R. D. Evans, *The Atomic Nucleus*, McGraw-Hill, New York (1955).
N. Feather, *Electricity and Matter*, Edinburgh University Press, Edinburgh (1968).
G. Gamow, *Atomic Energy in Cosmic and Human Life*, Cambridge University Press, Cambridge (1947).
H. G. Graetzer and D. L. Anderson, *The Discovery of Nuclear Fission*, Van Nostrand Reinhold, New York (1971).
O. Hahn, *New Atoms*, Elsevier, New York (1950).
O. Hahn, 'The Discovery of Fission', *Scientific American*, **198**, 76–84 (1958).
E. N. Hiebert, *The Impact of Atomic Energy*, Faith and Life Press, Newton (1961).
E. J. Holmyard, *Alchemy*, Penguin, Harmsworth (1957).
G. Holton, *Introduction to Concepts and Theories in Physical Science*, Addison-Wesley, Reading, Mass. (1952).
A. J. Hopkins, *Alchemy, Child of Greek Philosophy*, AMS Press, New York (1934) (reprinted 1967).
A. J. Ihde, *The Development of Modern Chemistry*, Harper & Row, New York (1964).
E. N. Jenkins, *Radioactivity: A Science in its Historical and Social Context*, Wykeham, London (1979).
H. Kopp, *Die Alchemie in älterer und neuerer Zeit*, C. Winter, Heidelberg (1886) (reprinted Hildesheim, Olms, 1962 and 1971).
E. Kremers and G. Urdang, *History of Pharmacy*, Lippincott, Philadelphia (1940) (4th revised edn by G. Sonnedecker, 1976).

Li Ch'iao-Ping, 'The Chemical Arts of Old China', *Journal of Chemical Education*, 25 (1948).

J. Lindsay, *The Origins of Alchemy in Graeco-Roman Egypt*, Barnes & Noble, New York (1970).

E. O. v. Lippmann, *Entstehung und Ausbreitung der Alchemie*, Springer, Berlin (1919) (includes 130 page appendix on history of the metals in antiquity) (reprinted Verlag Chemie, Weinheim, 1954).

E. v. Meyer, *A History of Chemistry from Earliest Times to the Present Day*, Macmillan, London (1891) (reprinted Arno Press, NY, 1975).

J. Needham, *Science and Civilisation in China*, Vol. 5: *Chemistry and Chemical Technology*, Cambridge University Press, Cambridge (1974).

J. R. Partington, *A Short History of Chemistry*, Macmillan, London (1937).

J. Read, *Prelude to Chemistry*, Macmillan, New York (1937) (London, 1936).

H. S. Redgrove, *Alchemy: Ancient and Modern*, Rider & Son, London (1922) (2nd edn University Books Inc., New York, 1969).

A. Romer, *The Discovery of Radioactivity and Transmutation*, Dover, New York (1964).

E. Rutherford, *The Newer Alchemy*, Cambridge University Press, Cambridge (1937).

G. T. Seaborg, *Transuranium Elements: Products of Modern Alchemy*, Dowden, Stroudsburg, Pa. (1978).

G. T. Seaborg, *The Transuranium Elements*, Yale University Press, New Haven (1958).

G. T. Seaborg, *Man-made Transuranium Elements*, Prentice-Hall, Englewood Cliffs (1963).

F. H. Schmidt and D. Bodansky, *The Energy Controversy: The Fight over Nuclear Power*, Albion, San Francisco (1977).

H. J. Sheppard, 'The Ouroboros and the Unity of Matter in Alchemy: A Study in Origins', *Ambix*, 10, 83–96 (1962).

H. J. Sheppard, 'Alchemy: Origin or Origins?', *Ambix*, 17, 69–84 (1970).

N. Sivin, *Chinese Alchemy*, Harvard University Press, Cambridge, Mass. (1968).

A. K. Smith, *A Peril and a Hope: The Scientists' Movement in America*, 1945–47, MIT Press, Cambridge, Mass. (1970).

H. D. Smyth, A General Account of the Development of Methods of Using Atomic Energy for Military Purposes under the Auspices of the United States Government 1940–45, U.S. Government, Washington, DC (1945).

F. Soddy, *Science and Life*, John Murray, London (1920).

F. Soddy, *The Story of Atomic Energy*, New Atlantis, London (1949).

J. W. van Spronsen, *The Periodic System of Chemical Elements*, Elsevier, Amsterdam (1969).

F. Strassman, *Kernspaltung: Berlin, Dezember 1938*, Mainz (1978).

F. S. Taylor, *The Alchemists*, Schuman, New York (1949) (reprinted Arno Press, New York, 1976).

T. J. Trenn, *Radioactivity and Atomic Theory*, Taylor & Francis, London (1975).

T. J. Trenn, *The Self-Splitting Atom*, Taylor & Francis, London (1977).

A. E. Waite, *Lives of Alchemystical Philosophers*, Redway, London (1888) (reprinted 1970).

S. R. Weart, *Scientists in Power*, Harvard University Press, Cambridge, Mass. (1979).

M. E. Weeks, *Discovery of the Elements*, 7th edn (H.M. Leicester, ed.), Chemical Education, Easton (1968).

R. R. Wilson and R. Littauer, *Accelerators: Machines of Nuclear Physics*, Doubleday, New York (1960).

A. Wyatt, *The Nuclear Challenge: Understanding the Debate*, The Book Press, Toronto (1978).

F. A. Yates, *The Rosicrucian Enlightenment*, Paladin, London (1972).

Subject Index

Fission, 1
  cross-section, 88
  fragments, 85
  products, 5, 85
  reactor, 97
Fixation, 49
Fluorescence, 58
Four-element theory, 31–32, 40, 53
  modified, 47, 49
Fractional crystallization, 28
French Academy of Science, 3
Fuel dividend, 102
Fusion, 103
  devices, 105–106
  reactor, 97
  toroids, 104
  transmutation, 103

Galenic medicine, 47
Gallium, 83
Gamma ray photons, 75
Gamma rays, 11, 20, 59, 76
Gas-discharge tube, 58–59, 79
Gaseous diffusion, 90
Geiger-Müller counter, 4, 20, 77
Gold, 32, 33, 40
Graphite, as moderator, 88
Greek philosophy, 31

'H' particles, 69–71
Half-life, 12, 16, 18, 22–23, 26–27, 73
Harran, 39
Helium, 59, 63
  nuclei, 59
Hermeticism, 52
Hermetic
  seal, 38
  vessels, 39
High energy neutrons, 79
Hybrid reactors, 106
Hydrocarbon reserves, 97, 103, 107
  competition for, 98
  conservation of, 68
  dependence upon, 103
Hydrogen, as nuclear constituent, 72–73
Hydropower, 99

Iatrochemistry, 45, 49, 108
Indian alchemy, 35

Induced radioactivity, 80
Inertial confinement, 105
Institut du Radium, 4
Instrumentation, importance of, 77
Ionization, 59
  current, 11
  by collision, 18
Iridium, 29
Irradiation
  by alpha particles, 70
  of uranium, 27
Isomeric
  isotopes, 22
  series, 28
Isotopes, 73
  mixture of, 22
  stable and unstable, 73, 80
Isotopic varieties, 5

Jupiter, 34

Knock-on recoil, 69
Krypton, 59, 85

Lanthanum, 85
  carrier, 23
  conundrum, 29, 83
Laser, 105
  pulse, 105
Later alchemy, 41
Latin alchemy, 40
Law
  of constant proportions, 54
  of radioactive change, 11, 15
Linear accelerator, 94
Long range alpha particles, 71
Luminescence, 59, 61

Macrocosm, 45
  –microcosm analogy, 32, 44, 51
'Mad-hatter' syndrome, 52
Magnesium, 21
Magnetic
  deviation of rays, 58
  'bottle', 105
Mars, 34
Mass
  spectrometry, 86
  defect, 86

Materia prima, 34, 44, 54–55, 57, 64,
71–72, 80, 108
Max-Planck-Gesellschaft, 7
Medicinal alchemy, 35, 40, 45, 53
Mercury, 33, 34, 40
    role of, 52
    theory of metals, 47
Mesothorium, 5, 23
Metabolon, 16
Metallic
    calx, 53
    perfection, 33, 41
Metallurgy, 35
Method as medium, 45, 64, 66, 72
Microcosm, 45
Microtron, 94
Moderator, 5, 78, 88, 106
Monomolecular chemical reaction, 15
Mystical alchemists, 50, 53

Natural
    reactor, 93
    transmutation, 56–57, 63–64
Near-breeder reactor, 100
Negative electron, 21
Neon, 59
Neptunium, 82–84, 89
Neutron, 3, 20, 75, 80, 87
    activation, 78, 83, 87
    as philosophers' stone, 75, 78–79, 84,
    93
    bombardment, 83
    capture, 82, 87–88
    discovery of, 76
    emission, 21
    energy, spectrum of, 93
    escape, 87
    excess, 85
    flux, 87–88, 100, 106
    flux, energy of, 78
    irradiation, 22–23
    life, 89
    multiplication constant, 88–89
    reflector, 90
'New gold', 56, 86, 96
    –energy, 72, 96, 108
'New pearl of great price', 41, 107

Newer alchemy, 56
    and older alchemy compared, 108
Nitrogen anomaly, 70–71, 87
Nobel Prize, 1
Nuclear
    chemists, 29
    constituents, 73
    energy, 109
    fission, 29, 56, 83–85, 99
    fission, proposed, 83
    isomerism, 7
    model of the atom, 2, 71
    power, 94, 99
    reaction, 78
    reaction, sustained, 88
    reactor, 1, 5, 88–89, 93
    stability, 80
    testing, control of, 94
    waste, 95
    weapons, 93–94, 106
Nucleus, 2, 109

Oklo phenomenon, 93
Older alchemy, 56
Origin of alchemy, 31
Osmium, 29
Ouroboros, 34–35, 104
Oxidation, 53

Pacing, of chain reaction, 89–90
Palladium, 29
Paraffin, 78
Particle accelerator, 79
Path of stability, 81–82, 85
Perfectibility, 44
Perfection, principle of, 30, 33, 36
Periodic table of the chemical elements,
    65, 82–83
Pharmacology, 47
Pharmacy, 45
Philosophers' stone, 30, 33, 38, 41, 44,
    47, 50, 56, 64, 72, 77, 86, 95, 108
Phlogiston, 53
Phosphorus, 21
Photofission, 90, 92
Photographic effects, 59
Photon, 4
Physical alchemy, 53

Name Index

Abelson, P., 83
Albert the Great, 40–41
Amaldi, E., 78
Anderson, C.D., 76
Arnold of Villanova, 40–41, 45
Aston, F.W., 86
Bacon, R., 40–41, 45
Becher, J.J., 53
Becker, H., 4, 75, 77
Becquerel, A.H., 13–14, 58–59, 61, 80
Berzelius, J.J., 55
Blackett, P.M.S., 76
Boehme, J., 50
Bohr, N., 2, 69–70
Bothe, W., 4, 75, 77
Boyle, R., 53–54

Cameron, A. T., 66–67
Chadwick, J., 75–76
Cockroft, J.D., 79
Crookes, W., 57, 61, 82
Curie, M., 13–15, 29, 59, 61, 79
Curie, P., 59, 61

Dalton, J., 54
Darwin, C.G., 69
Dirac, P.A.M., 76
Dumas, J.B.A., 55

Einstein, A., 85

Fajans, K., 82
Fermi, E., 5, 78, 80, 85

Fludd, R., 52
Frisch, O.R., 85

Galen, 47
Geber, 40, 47
Giesel, F., 14, 59, 63
Glauber, J.R., 49

Hahn, O., 1, 5–7, 22, 26, 29, 84–85, 106
Halban, H. von, 5
Helmont, J.B. van, 49–51
Helvetius, 50
Hermes, 38
Hermes Trismegistos, 31, 39
Hofmann, K. A., 14
Joliot, J.-F., 1, 4–5, 20, 75, 77, 80
Joliot-Curie, I., 1, 4–5, 7, 20, 29, 75, 77, 83

Kelvin (Sir William Thomson), 61
Ko Hung, 36, 38
Kowarski, L., 5

Langevin, P., 4
Lavoisier, A.L., 53
Lawrence, E.O., 79
Libavius, A., 49
Livingstone, 79
Lockyer, J.N., 19
Lully, R., 40–41

127